Mobile App Development

Mobile App Development 101:

A Step-by-Step Guide for Beginners

Noah Bailey

Table of Contents

INTRODUCTION

Welcome to " Mobile App Development: Mobile App Development 101- A Step-by-Step Guide for Beginners." Mobile applications have become essential to our everyday lives in a time when interaction with screens comes mostly through touch and swipe gestures. Mobile applications, ranging from social networking to productivity tools and games to health apps, have completely changed how we engage with the digital world.

Have you ever wondered, "How do they create something like this?" when browsing your favorite app? Or you've always wanted to develop your app idea but didn't know where to start. You are in the proper place if that is the case.

You, the aspiring app developer with little to no expertise, are the target audience for this book. Mobile app development is an intricate and demanding industry with its technical terminology and intricate coding languages. Nonetheless, anyone can learn to design mobile apps with the correct direction and a systematic approach.

In the following pages, we'll take you on a journey from complete novice to self-assured mobile app developer. Regardless of your interest in developing cross-platform apps or apps for iOS and Android, this thorough tutorial will provide you with the information and abilities you need to get started.

In this book, we'll simplify the complex process of developing mobile apps by breaking it into digestible chunks. We'll walk you through the fundamentals of programming, help you set up your development environment, educate you on creating intuitive user interfaces, and demonstrate how to utilize code to realize your app concepts.

However, it extends beyond the technical aspects of development in this book. The best practices for user experience, app store optimization, and app design will also be covered. You'll discover how to test your apps, submit them to app stores, and even think about monetization tactics.

Providing you with practical skills is equally important as providing a theoretical understanding. To make sure you can put what you learn into practice, we'll give you activities, examples, and real-world perspectives.

This book is primarily meant for those who have never developed a mobile app but are interested in the field. You can find helpful information here whether you're a student looking to change careers, an entrepreneur with a great app concept, or a tech enthusiast.

By the end of this journey, regardless of your experience, you will have acquired the necessary skills and knowledge to establish your mobile apps from scratch, transforming your concepts into functional and engaging digital realities.

Let's get started if you're prepared to take on this thrilling journey into the realm of mobile app development. Now is the moment to develop your concepts into applications that have the potential to impact individuals positively" lives everywhere. Welcome to "Mobile App Development: Mobile App Development 101- A Step-by-Step Guide for Beginners." Let's begin your journey today.

CHAPTER I

Understanding Mobile App Development

What is mobile app development?

The art and science of developing software applications for mobile devices, particularly tablets and smartphones, is known as mobile app development. These applications have become ever-present in the modern digital landscape, shaping how we communicate, work, play, and even shop. They have revolutionized our daily lives, putting a world of information and functionality right at our fingertips.

At its core, mobile app development is about crafting software tailored to mobile devices' unique capabilities and constraints. It involves a blend of creativity, problem-solving, and technical expertise. The goal is to provide users with a seamless and engaging experience while addressing their specific needs and desires.

To understand mobile app development better, it's crucial to grasp its key components. Firstly, the choice of platform is essential. Developers must decide whether to create apps for the Android ecosystem, iOS (Apple's mobile operating system), or both. Each platform has its own set of development tools, programming languages, and design guidelines, which require distinct approaches to app creation.

Programming languages are the foundation of app development. Java and Kotlin are commonly used for Android, while iOS developers primarily work with Swift or Objective-C. These languages allow developers to write the code that powers the app's functionality. Code can range from simple commands to complex algorithms, all working together to deliver the desired user experience.

The user interface (UI for short) and the user experience, or UX for short, design are equally important. UI focuses on the app's visual elements, including layout, colors, and interactive features like buttons and menus. UX, conversely, is concerned with how users interact with the app and the overall satisfaction they derive from the experience. Effective UI/UX design is crucial for keeping users engaged and ensuring they can navigate the app effortlessly.

Mobile app development doesn't occur in isolation; it's a collaborative process often involving multiple team members. These teams typically comprise developers, designers, testers, and project managers, each playing a unique role in bringing the app to life. Developers write the code, designers create the visuals, testers identify and fix bugs, and project managers ensure that the development process stays on track and aligns with the project's goals.

Testing is a critical phase in app development. Developers rigorously test their apps to identify and eliminate any issues that could affect the user experience. This process includes functional, usability, and performance testing to guarantee the app works as intended and runs smoothly on various devices.
Once the app is complete and thoroughly tested, it's time to prepare for its release to the world. This entails

following the guidelines and requirements set forth by the individual app stores, such as for iOS, the Apple App Store and for Android, the Google Play Store. Meeting these standards is essential to ensure the app reaches its intended audience and can be easily discovered by potential users.

In summary, mobile app development is a dynamic and ever-evolving field that empowers individuals and organizations to create digital solutions that cater to the diverse needs of mobile device users. It encompasses a range of skills, from coding and design to testing and deployment, and is driven by a commitment to delivering high-quality, user-centric experiences. As mobile devices continue to shape our world, mobile app development remains at the forefront, enabling innovation and convenience in an increasingly digital age.

Why is mobile app development important?

In today's digitally driven world, mobile app development is a cornerstone of innovation and convenience. It has emerged as a transformative force, reshaping how we interact, communicate, and conduct business. Mobile applications, running on smartphones and tablets, have become indispensable tools that bridge the gap between our needs and technology's solutions. In this section, we delve into why mobile app development is important, exploring its multifaceted impact on individuals, businesses, and society at large.

Mobile app development addresses the ever-expanding demands of modern life. Users can access many services that cater to diverse needs with a few taps on a screen. Whether it's ordering food, booking a ride, tracking fitness goals, or managing personal finances, there's an app for nearly everything. This convenience is pivotal in simplifying daily routines, saving time, and enhancing overall productivity. The ability to have a personalized digital assistant readily available has elevated the importance of mobile apps in our lives.

Mobile apps are not just tools for personal efficiency but also instruments of connectivity. The way we communicate to each other and share our lives has changed a lot because of social media sites which include Facebook, Instagram, and Twitter. Messaging apps like WhatsApp and Telegram have transformed communication into instant, global exchanges. Even in business, apps like Slack have redefined workplace collaboration. Mobile app development has thus deepened human connections, making the world feel smaller and more interconnected.

Mobile app development is a critical avenue for growth and innovation for businesses. Companies now recognize the immense potential of reaching customers through their smartphones. Mobile apps provide a direct and personalized channel to engage with users, showcase products or services, and facilitate transactions. E-commerce giants like Amazon and Alibaba have capitalized on this trend, offering seamless shopping experiences through their apps. Also, mobile apps let businesses collect useful information about how users act and what they like, which can help guide strategic choices and marketing efforts.

In addition to established businesses, mobile app development has nurtured a fertile ground for startups and entrepreneurs. The low barrier to entry and the ability to target niche markets have encouraged innovation in various sectors. The success stories of startups like Uber and Airbnb, which disrupted traditional industries through mobile apps, serve as a testament to the transformative power of app development.

Mobile app development has left no sector untouched. It has empowered industries ranging from healthcare to education, entertainment to finance, and transportation to agriculture. In healthcare, apps allow patients to monitor their health, schedule appointments, and even consult with doctors remotely. Educational apps provide access to a world of knowledge, making learning accessible to all, regardless of geographic location. Streaming services like Netflix and Spotify have redefined the entertainment industry, while mobile banking apps have revolutionized how we manage finances.

Moreover, mobile app development has fostered the growth of specialized services. Dating apps like Tinder have changed how people meet and form relationships. Travel apps like Airbnb make it possible to find unique accommodations worldwide. Food delivery apps like Uber Eats have transformed dining experiences, and ride-sharing apps like Lyft have disrupted the transportation industry.

A key reason for the importance of mobile app development lies in its ability to offer a superior user experience. Well-designed apps provide intuitive interfaces, smooth navigation, and responsive performance. This focus on user experience is crucial because it directly impacts user retention and engagement. Users are more likely to continue utilizing an app that offers a pleasant and efficient experience. Conversely, apps with poor design or functionality often get uninstalled quickly. As a result, businesses and developers alike prioritize user-centric design and continuous improvement to stay competitive in the app market.

The global reach of mobile app development is unprecedented. With a single app available in app stores, developers can potentially reach billions of users worldwide. This level of accessibility and scalability is unparalleled in the history of technology. It allows developers to transcend geographical boundaries and cultural barriers, enabling their creations to impact a diverse and global audience.

Moreover, mobile apps have opened doors for inclusivity and accessibility. They have made technology more approachable for individuals with disabilities, providing features like screen readers, voice commands, and other assistive technologies. This accessibility underscores the importance of mobile app development in creating a more inclusive digital world.

Mobile app development also offers lucrative opportunities for developers and businesses. While many apps are offered for free, they often incorporate monetization strategies such as in-app advertising, subscription models, or one-time purchases. This diversity in revenue streams has made app development a viable and sustainable career path for many individuals and a profitable venture for businesses. Successful apps can generate substantial income, making them attractive investments.

Competition is a driving force in the world of mobile app development. The presence of multiple apps offering similar services pushes developers to innovate continually. They seek ways to differentiate their products through unique features, improved user interfaces, and better performance. This competition benefits users, leading to higher-quality apps and broader choices.

Furthermore, the app stores, such as the Apple App Store and Google Play Store, play a crucial role in fostering innovation. They provide a platform for developers to showcase their creations, making it easier for users to discover new and innovative apps. The recognition and exposure offered by app stores can catapult a developer or business to success.

The importance of mobile app development is closely tied to its ability to adapt to evolving lifestyles and technological trends. As our lives become increasingly fast-paced and interconnected, mobile apps serve as agile solutions that cater to changing needs. They develop alongside emerging technologies, such as augmented reality (AR) and virtual reality (VR), voice assistants, and the Internet of Things (IoT). This adaptability ensures that mobile app development remains relevant and indispensable in an ever-shifting digital landscape.

In conclusion, mobile app development has become integral to our lives, transforming how we live, work, and connect with others. Its importance is evident in its ability to simplify everyday tasks, enhance communication, drive business growth, empower industries, and enrich user experiences. As technology advances, mobile app development will continue to evolve, shaping our digital future and expanding the horizons of what is possible in the palm of our hands. Its significance is not just in the apps themselves but in their profound impact on our world.

Different platforms (iOS, Android, Cross-Platform)

In the ever-evolving landscape of mobile app development, the choice of platform holds paramount importance. Aspiring app developers and businesses embarking on the journey to create mobile applications must decide between various platforms, the most prominent being iOS, Android, and cross-platform development. Each platform presents unique opportunities and challenges, and selecting the right one is a pivotal decision that influences the app's reach, functionality, and user experience. In this section, we delve into the characteristics, advantages, and considerations of iOS, Android, and cross-platform development, enabling a comprehensive understanding of the platform landscape and aiding in informed decision-making.

iOS, Apple's mobile operating system, is renowned for its premium user experience, polished interface, and strong focus on security and privacy. iCloud, Apple Music, the App Store, and other services are all part of the larger Apple ecosystem, and iOS devices—which include the iPhone and iPad—are renowned for their smooth design and seamless connectivity. iOS app development typically uses Swift or Objective-C, Apple's official programming languages.

One of the key advantages of developing for iOS is the consistency of the user experience. Apple tightly controls its devices' hardware and software aspects, resulting in a predictable and standardized environment for app developers. This control extends to the App Store, where stringent review processes ensure the quality and security of apps available to users. The App Store is also known for its higher average revenue per user compared to other platforms, making it an attractive choice for monetization.

Moreover, iOS users adopt new software updates swiftly, simplifying the task of maintaining compatibility with the latest devices and features. This phenomenon, known as the "iOS ecosystem advantage," allows developers to target a relatively homogeneous user base.

However, iOS development has its share of challenges. The development environment is exclusively macOS-based, which can be a barrier for developers using other operating systems. Additionally, the approval process for getting apps into the App Store can be rigorous, with the possibility of rejection for violating Apple's guidelines. Developing for iOS also means limited market reach, as iPhones and iPads make up a smaller share of the global device market compared to Android.

In stark contrast to iOS, the Android platform offers a diverse and fragmented ecosystem. Android, developed by Google, is an open-source operating system used by numerous manufacturers across various devices, from flagship smartphones to budget tablets and even smart TVs and wearables. This diversity is both a strength and a challenge for Android app developers.

Android's open nature allows for greater flexibility and customization, making it an attractive choice for device manufacturers. However, this diversity also poses challenges when it comes to app development. Developers must contend with many screen sizes, resolutions, hardware capabilities, and operating system versions. This fragmentation can result to compatibility issues and requires thorough testing on various devices. Two of the most popular programming languages used to develop Android apps are Java and Kotlin. Kotlin, in particular, has gained popularity for its conciseness and expressiveness, making development more efficient. Android's app distribution model differs from iOS's. Rather than a single centralized store, Android apps can be distributed through various app stores and even directly from a developer's website. This flexibility can be advantageous for reaching a global audience, but it also means that the Android ecosystem is more susceptible to security risks and potentially harmful apps.

Monetization options for Android apps are diverse, with developers having the choice of advertising, in-app purchases, subscriptions, or one-time purchases. While Android's market share is more significant than iOS's in terms of global device usage, it tends to generate lower average revenue per user. Therefore, developers often rely on a larger user base to achieve their revenue goals.

Cross-platform development has gained prominence as a solution to the challenges posed by the iOS and Android divide. By using this method, developers can cut down on development time and effort by producing a single codebase that can be used on several platforms. Cross-platform frameworks that are in demand include PhoneGap, Xamarin, Flutter, and React Native.

React Native, developed by Facebook, has gained widespread adoption due to its ability to produce near-native performance and its large and active community. It enable developers to write code in JavaScript or TypeScript while providing access to native platform features.

Flutter, created by Google, is another notable cross-platform framework known for its fast development cycles and expressive user interfaces. It uses the Dart programming language and compiles to native code, resulting in high-performance apps.

Xamarin, now a part of Microsoft, allows developers to use C# to build cross-platform apps. It provides deep integration with the native platform and access to various libraries.

PhoneGap, based on web technologies like HTML, CSS, and JavaScript, offers a straightforward way to create cross-platform apps. It leverages a webview to render the app's user interface, making it accessible to web developers.

Cross-platform development offers several advantages. By enabling developers to write code only once and deliver it across numerous platforms, it streamlines development processes and saves time and cost. This approach also addresses the challenge of reaching a broad audience by enabling simultaneous development for iOS and Android. Moreover, cross-platform frameworks often provide a rich set of pre-built components and plugins, further expediting development.

However, cross-platform development has its limitations. While these frameworks strive for near-native performance, there may still be differences in performance and user experience compared to fully native apps. Additionally, access to platform-specific features and APIs may be limited or require additional customization. Finally, cross-platform apps may lag when it comes to adopting the latest platform-specific features and design guidelines.

Choosing a mobile app development platform ultimately depends on numerous factors, such as the target audience, budget, development resources, and project goals.

For those aiming to create polished, high-performance apps that offer a premium user experience and are willing to invest in the Apple ecosystem, iOS development is an excellent choice. iOS apps are particularly favored by finance, healthcare, and media industries, where security and user experience are paramount.

On the other hand, Android caters to a broader, more diverse user base and can be the preferred platform for businesses looking to reach a wide audience. Android is especially suitable for customization and hardware integration apps, such as IoT applications.

Cross-platform development is a compelling option for developers and businesses seeking efficiency and cost-effectiveness without sacrificing quality. Cross-platform frameworks have matured significantly in recent years, making achieving near-native performance and maintaining a single codebase possible. This approach particularly appeals to startups and small to medium-sized businesses with limited resources.

In practice, some projects may even benefit from a combination of native and cross-platform development. For example, critical core features of an app could be developed natively for iOS and Android to ensure optimal performance, while less critical components or features could be implemented using cross-platform technologies.

It's worth noting that the mobile app development landscape is dynamic, and new tools and technologies continually emerge. Therefore, keeping informed about the latest developments and industry trends is essential when making platform decisions.

The selection of a mobile app development platform is a critical decision that profoundly influences an app's success. iOS, with its focus on user experience and security, offers a premium environment favored by users in affluent markets. Android, known for its diversity, provides access to a vast user base but demands careful consideration of fragmentation and compatibility challenges. While efficient and cost-effective, cross-platform development may require trade-offs in terms of performance and access to platform-specific features. Ultimately, the decision should align with project objectives, target audience, and available resources. The platform chosen should empower developers to create apps that resonate with users, whether by delivering a seamless and immersive experience on iOS, reaching a diverse audience on Android, or efficiently developing for multiple platforms with cross-platform technologies. The key to success lies in deciding on the right platform and the skillful execution of the development process, ensuring that the app meets user needs and stands out in an increasingly competitive app market. As technology advances, the choices available to developers will expand, offering new opportunities and challenges in the dynamic world of mobile app development.

Tools and technologies utilized in mobile app development

Mobile app development has emerged as a dynamic and rapidly evolving field, reshaping how we interact with technology daily. Behind the scenes of every mobile app lies a complex framework of tools and technologies that enable developers to turn their ideas into interactive, functional realities. This section delves into the ecosystem of tools and technologies used in mobile app development, providing insights into the components that power our favorite apps and drive innovation in the digital age.

At the heart of mobile app development are Integrated Development Environments (IDEs). These software applications provide a comprehensive environment for writing, testing, and debugging code. For iOS development, Xcode is the official IDE, while Android developers primarily use Android Studio. These IDEs offer a range of features, including code editors, debugging tools, emulators, and access to libraries and APIs. They simplify the development process by providing a unified interface for designing, coding, and testing apps.

The fundamental building blocks for developing mobile apps are programming languages. Different platforms use different languages. For iOS development, Swift and Objective-C are the primary languages. Swift, introduced by Apple in 2014, has gained popularity for its readability and performance. Objective-C, although older, is still used in legacy projects. Java was traditionally the primary language on the Android side, but Kotlin has gained traction due to its conciseness and expressiveness. Depending on the chosen framework, cross-platform development often leverages JavaScript, TypeScript, Dart, or C#.

Creating a compelling User Interface (UI)/User Experience (UX) is crucial in mobile app development. Design tools like Sketch, Adobe XD, and Figma are commonly used to create wireframes, prototypes, and mockups. These tools enable designers and developers to collaborate on the visual aspects of the app, ensuring a user-friendly and aesthetically pleasing interface. Additionally, UI frameworks and libraries, such as UIKit for iOS and Android's Material Design, provide pre-built components and design guidelines for consistent and visually appealing app interfaces.

Cross-platform development frameworks have gained prominence for their ability to streamline development efforts across multiple platforms. These frameworks let developers write code once and deploy it on iOS and Android, reducing development time and effort. Popular frameworks include React Native, Flutter, Xamarin, and PhoneGap. React Native, for example, uses JavaScript and React to build near-native apps. Flutter, created by Google, uses the Dart language and offers fast development cycles and expressive UIs. Xamarin, part of Microsoft, leverages C# to develop cross-platform apps, while PhoneGap relies on web technologies like HTML, CSS, and JavaScript.

Application Programming Interfaces (APIs) enable communication between a mobile app and external services or data sources. They provide a standardized way for apps to access features like geolocation, camera, push notifications, and social media integration. Services like Google Maps API, Facebook Graph API, and Firebase Cloud Messaging (FCM) are commonly used to enhance functionality and user engagement in mobile app development. RESTful APIs and GraphQL are popular choices for web-based communication between mobile apps and servers.

Version control systems, such as Git, are crucial for tracking changes to code and collaborating with team members. Git allows developers to work on different branches of the codebase simultaneously, merge changes, and roll back to previous versions if needed. Platforms like GitHub and Bitbucket provide hosting services for Git repositories, enabling collaboration and code sharing among developers.

Storing and managing data is a fundamental aspect of mobile app development. Database management systems (DBMS) like SQLite, MySQL, PostgreSQL, and Firebase Realtime Database facilitate data storage and retrieval. SQLite, a lightweight and embedded DBMS, is commonly used in mobile app development due to its efficiency and portability. Firebase Realtime Database, on the other hand, is a cloud-based NoSQL database that offers real-time synchronization, making it suitable for apps requiring live updates.

Continuous Integration and Continuous Deployment (also referred to as CI/CD) pipelines automate the process of building, testing, as well as deploying mobile apps. Tools like Jenkins, Travis CI, and CircleCI are employed to ensure that code changes are thoroughly tested and integrated into the app's development or production environment. CI/CD pipelines help maintain code quality, reduce manual errors, and enable rapid delivery of app updates to users.

Understanding user behavior and app performance is essential for app optimization. Tools like Google Analytics for Mobile Apps, Firebase Analytics, and AppDynamics provide insights into user engagement, app crashes, and performance bottlenecks. These analytics tools help developers identify areas for improvement and prioritize feature enhancements.

Bringing an app to users' devices requires careful consideration of distribution channels. Apple App and

Google Play Store are the primary platforms for app distribution on iOS and Android, respectively. These stores have guidelines and review processes that apps must adhere to before publishing. Additionally, tools like Apple's TestFlight and Google Play's Internal Testing allow developers to test beta versions of their apps with a select group of users before a wider release.

Security is a top priority in mobile app development. Tools and technologies like SSL/TLS encryption, certificate pinning, and secure coding practices are employed to protect user data and ensure the app's integrity. Security scanning tools and penetration testing help identify vulnerabilities that malicious actors could exploit.

The advent of Augmented Reality (or AR) and Virtual Reality (or VR) has opened new possibilities in mobile app development. ARKit for iOS and ARCore for Android provide frameworks for creating augmented reality experiences. VR development is supported through platforms like Google VR and Oculus. These technologies are transforming gaming, education, healthcare, and real estate industries by offering immersive and interactive experiences.

The tools and technologies used in mobile app development constitute a vibrant and evolving ecosystem. Developers and businesses must navigate this landscape to create apps that meet user needs, deliver outstanding user experiences, and stay competitive in a rapidly changing digital world. The selection of tools and technologies depends on project requirements, goals, and target platforms. Whether developing for iOS, Android, or both, the right combination of tools and technologies empowers developers to turn their visions into successful and impactful mobile applications. As technology advances, the mobile app development ecosystem will continue to expand, offering new possibilities and challenges for developers and users alike.

CHAPTER II

Setting Up Your Development Environment

Choosing the right operating system (Windows, macOS, Linux)

The choice of an operating system (OS) is a fundamental decision that profoundly influences how we interact with computers and the digital world. Among the plethora of operating systems available, three major players dominate the landscape: Windows, macOS, and Linux. Each OS offers a distinct set of features, capabilities, and philosophies, making the decision critical, whether for personal use, professional tasks, or software development. This section explores the considerations of choosing the right operating system and the unique characteristics of Windows, macOS, and Linux.

The operating system is the foundational software layer that manages computer hardware and allows users and applications to interact with the machine. It influences the user interface, software compatibility, security, and overall user experience. As such, selecting the right operating system is a decision that warrants careful thought and consideration.

Microsoft Windows, commonly called Windows, has long been the dominant operating system for personal computers. It is known for its widespread adoption in homes and businesses, making it the go-to choice for many users. Windows provides a user-friendly interface that has evolved over several iterations, including the popular Windows 7, 8, and more recent Windows 10 and 11.

One of the standout features of Windows is its extensive software library. Windows users enjoy access to a vast ecosystem of applications and games, ranging from productivity tools to entertainment software. Many popular software titles and games are developed primarily for the Windows platform, making it an attractive choice for casual and professional users.

Windows is also known for its compatibility with various hardware devices. It supports many peripherals, making it a versatile choice for users who require specialized hardware configurations or engage in tasks that demand broad hardware compatibility.

Windows often remains the preferred choice for businesses due to its extensive support for enterprise-level applications and a robust ecosystem of IT management tools. Additionally, Microsoft's suite of productivity software, including Microsoft Office, is optimized for Windows, making it the natural choice for organizations that rely on these tools.

However, Windows has its drawbacks. It is historically known for security vulnerabilities and is a frequent target for malware and viruses. Microsoft has made significant strides in improving security, but users must remain vigilant and employ additional security measures. Additionally, Windows is a paid operating system, and users often face the prospect of upgrading to newer versions, which can incur additional costs.

macOS, developed by Apple Inc., is the operating system that powers Apple's line of Macintosh computers. Known for its elegant design, seamless integration with Apple hardware, and loyal user base, macOS offers a unique computing experience. macOS has undergone several iterations, with names like Yosemite, El Capitan, and more recently, Big Sur and Monterey.

One of the standout features of macOS is its user interface, characterized by a clean, minimalist design. The Mac App Store provides a curated selection of software applications, ensuring a certain level of quality and security. The macOS interface also boasts features like Mission Control, which simplifies the management of multiple open applications and desktops.

Integrating macOS and other Apple devices, such as iPhones and iPads, is a significant selling point. Features like Handoff allow users to seamlessly transition tasks between their Mac and iOS devices, enhancing productivity and convenience.

macOS is lauded for its robust security features. It benefits from Apple's proactive approach to security, with features like Gatekeeper, FileVault, and built-in antivirus protections. The closed nature of the macOS ecosystem contributes to its security, as software installations are typically restricted to the App Store, reducing the risk of malware.

For creative professionals, macOS is often the preferred choice. It is renowned for its performance in innovative software applications like Adobe Creative Cloud and Final Cut Pro. The high-resolution Retina displays on MacBooks enhance the visual experience, making it ideal for graphic design, video editing, and content creation.

However, macOS comes with a premium price tag, as it is only available on Apple's hardware. This limits choices for users who prefer a wider selection of hardware options. Additionally, while macOS offers a robust selection of software, it may not have the same breadth and variety as the Windows ecosystem. Compatibility with specific enterprise-level software and specialized applications can be a consideration for business users.

Linux stands apart from Windows and macOS as an open-source operating system with a rich ecosystem of distributions (distros). Unlike Windows and macOS, Linux is not tied to a single company; instead, it is developed collaboratively by a global community of volunteers and organizations. Linux distributions like Ubuntu, Fedora, and Debian offer distinct user experiences and cater to various needs.

The open-source nature of Linux means it is freely available, making it an attractive choice for users who prioritize cost savings. Linux can be installed on various hardware, including older machines, revitalizing outdated computers and reducing electronic waste.

One of Linux's most significant strengths is its flexibility and customization. Users can choose from many desktop environments, each with its own look and feel. This flexibility extends to the choice of software, as Linux offers various applications for productivity, gaming, development, and more.

Linux is renowned for its stability and security. It is considered one of the most secure operating systems, with a robust permissions system and regular security updates. Additionally, Linux servers power a significant portion of the internet, attesting to its reliability and scalability.

For software developers, Linux is appealing due to its support for a broad range of programming languages and development tools. The command-line interface, a staple of Linux, provides powerful scripting and automation capabilities, making it a preferred platform for many developers.

However, Linux does have its challenges. For users used to Windows or macOS, the learning curve can be quite high, especially for those who are not experienced with the command line. Compatibility with specific proprietary software and hardware drivers can be an issue, although the situation has improved over the years. Depending on the chosen distribution, support for specific enterprise-level applications may also be limited on Linux.

In conclusion, the choice of an operating system ultimately depends on individual preferences, requirements, and priorities. Windows offers a vast software library, broad hardware compatibility, and familiarity for many users. macOS appeals to those who value design, security, and seamless integration with other Apple devices. With its open-source nature, customization options, and cost-efficiency, Linux is an excellent choice for those seeking flexibility and control.

The decision may also be influenced by the specific tasks a user or organization needs to accomplish. Windows excels in enterprise environments and gaming, macOS in creative and design fields, and Linux in development and customization. It is noteworthy that certain users choose to use virtualization or dual-booting in order to run several different operating systems on a single computer.

Ultimately, the right operating system is the one that aligns with an individual's or organization's goals, preferences, and workflow. The lines between these operating systems may blur as technology evolves, with features and capabilities converging. Nonetheless, the diverse choices in the operating system landscape ensure that users can find a platform that suits their unique needs and preferences.

Installing necessary software (IDE, SDKs)

In software development, installing essential software tools is the first step toward transforming ideas into functional applications. These tools form the backbone of the development process, providing programmers the means to write, test, and deploy code efficiently.

Integrated Development Environments (IDEs) and Software Development Kits (SDKs) are among the most critical components. In this section, we explore the significance of these tools, their role in software development, and considerations for selecting the right ones.

Integrated Development Environments (IDEs) are comprehensive software applications that streamline software development. They are the central hub where developers write, edit, debug, and manage code. A compelling IDE offers a range of features that simplify and enhance the coding experience.

One of the primary advantages of using an IDE is the convenience of a unified workspace. In contrast to using separate text editors, compilers, and debugging tools, an IDE combines all these functions into a single interface. This integration saves time and reduces the complexity of managing multiple software applications.

IDEs are equipped with code editors that offer syntax highlighting, code completion, and auto-formatting, among other features. These tools help developers write clean, error-free code more efficiently. Furthermore, IDEs provide powerful debugging capabilities, allowing developers to identify and rectify issues in their code, step through program execution, and inspect variables.

Another crucial aspect of IDEs is their support for version control systems like Git. With version control, developers can work together as a team, monitor changes made to their code, and go back to previous versions as needed. This is especially valuable in larger development projects.

IDEs are highly customizable, allowing developers to install plugins and extensions that enhance functionality. These plugins can range from language-specific tools to third-party services and frameworks integrations. The ability to tailor the development environment to specific project needs contributes to increased productivity.

Different programming languages often have dedicated IDEs that cater to their unique requirements. For example, Visual Studio is a popular choice for C# developers, while PyCharm is tailored for Python development. However, some IDEs, like Visual Studio Code, are versatile and support a wide range of programming languages, making them popular for developers working with multiple technologies.

When selecting an IDE, developers should consider factors such as language support, ease of use, debugging capabilities, and the availability of relevant plugins. Ultimately, the right IDE can significantly impact a developer's efficiency and the software's overall quality.

Software Development Kits (SDKs) are collections of software tools, libraries, and documentation that enable developers to create applications for specific platforms, frameworks, or technologies. SDKs provide the essential building blocks for implementing particular functionality, whether it's developing mobile apps, integrating with hardware devices, or working with cloud services.

One of the most well-known SDKs is the Android SDK, which empowers developers to build Android mobile operating system applications. It includes essential tools like the Android Debug Bridge (ADB) for device communication, user interface design libraries, and app development guidelines documentation.

Similarly, the iOS SDK equips developers with the tools and resources required to create applications for Apple's iOS platform. It includes Xcode, the official IDE for iOS development, as well as frameworks for user interface design, device features like geolocation, and integration with Apple services like iCloud and Siri.

SDKs extend beyond mobile app development. For example, the Amazon Web Services (AWS) SDK allows developers to interact with AWS services, enabling them to build scalable and cloud-powered applications. The Microsoft Azure SDK serves a similar purpose, offering tools for developing, deploying, and managing applications on the Azure cloud platform.

The functionality provided by an SDK varies depending on its purpose. Some SDKs focus on hardware integration, allowing developers to interact with sensors, cameras, and other device features. Others concentrate on web services, enabling developers to access APIs for data retrieval, authentication, and more.

When choosing an SDK, developers should consider their project requirements and the platform or service they intend to work with. An effective SDK should provide comprehensive documentation, code samples, and a community of developers who can offer support and guidance.

The world of software development is in a constant state of evolution. New tools and technologies emerge regularly, offering developers more options and capabilities. For example, serverless frameworks like AWS Lambda and Azure Functions have gained popularity for building event-driven applications without the need to manage servers. Additionally, containerization tools like Docker and orchestration platforms like Kubernetes have transformed how applications are deployed and scaled.

Furthermore, cloud-based development environments have become increasingly prevalent. These platforms, such as GitHub Codespaces and GitLab's Web IDE, enable developers to code, collaborate, and build software entirely in the cloud. This trend reflects a growing shift toward remote and distributed development teams.

In conclusion, installing necessary software tools, including Integrated Development Environments (IDEs) and Software Development Kits (SDKs), is a foundational step in the software development process. IDEs offer a unified workspace for coding, debugging, and managing projects, enhancing developer productivity. On the other hand, SDKs provide the essential building blocks and resources needed to create applications for specific platforms, frameworks, or technologies.

The choice of IDE and SDK should align with the development goals, programming languages, and platforms involved in a project. When selecting an IDE, developers should consider factors such as language support, debugging capabilities, ease of use, and available plugins. For SDKs, compatibility with the target platform, comprehensive documentation, and community support are critical considerations.

As technology advances, developers must stay informed about the latest tools and trends in the ever-evolving software development landscape. The right combination of IDEs and SDKs empowers developers to bring their software ideas to life efficiently and effectively, contributing to innovation and progress in the field of technology.

Setting up a development device or emulator

In software development, setting up a development device or emulator is a critical step that lays the foundation for creating and testing applications. Whether you're building mobile apps, web applications, or software for specific hardware, having the right development environment is essential for a seamless and efficient workflow. In this section, we delve into the significance of configuring a development device or emulator, exploring the nuances of each option and their implications for developers.

A development device refers to a physical piece of hardware on which software applications are developed, tested, and debugged. These devices range from personal computers and laptops to smartphones, tablets, and specialized hardware like IoT (Internet of Things) devices. The choice of a development device depends on the target platform and the nature of the software being developed.

A personal computer or laptop serves as the primary development device for web development and software that runs on desktop operating systems like Windows, macOS, or Linux. Developers typically set up their development environment by installing the necessary tools, including Integrated Development Environments (IDEs), compilers, code editors, and software libraries.

A physical smartphone or tablet is often required when developing mobile applications for platforms like Android or iOS. These devices allow developers to test their applications in a real-world environment, ensuring that the user experience and functionality meet expectations. The setup process involves enabling developer options, connecting the device to the development machine, and configuring it for debugging and testing.

In the case of IoT development, specialized hardware devices are used as development devices. These devices may include microcontrollers, sensors, and actuators, each serving a unique role in the IoT ecosystem. Developers connect to these devices via USB, Wi-Fi, or other communication protocols, using development kits and tools provided by the device manufacturer.

Development devices offer several advantages. They provide a realistic testing environment, allowing developers to assess their software's performance on actual hardware. Developers can observe how the software interacts with device-specific features, sensors, and screen sizes. Additionally, development devices enable testing under real-world conditions, such as network connectivity variations and hardware limitations.

However, using development devices can also pose challenges. Acquiring and maintaining various devices for testing can be costly and time-consuming, mainly when supporting multiple platforms and device models. Developers must also consider compatibility issues when targeting different hardware configurations.

Emulators, or virtual machines or simulators, offer an alternative to physical development devices. These software-based tools replicate the behavior of a specific platform, device, or operating system, allowing developers to test their applications without needing physical hardware. Emulators are particularly valuable when developing for mobile platforms, gaming consoles, or legacy hardware.

Mobile app development, in particular, benefits from the use of emulators. Android emulators like the Android Virtual Device (AVD) and iOS simulators provided by Xcode simulate the behavior of mobile devices on a developer's computer. These emulators mimic device characteristics such as screen size, resolution, and hardware features, enabling developers to test their apps under various conditions.

Emulators are versatile tools that offer several advantages. They eliminate the need for physical devices, reducing costs and simplifying device management. Emulators also facilitate rapid testing and debugging, allowing developers to run, pause, and inspect code easily. Furthermore, they offer features like screen recording and the ability to simulate various device configurations, making them valuable for app development and user interface testing.

Emulators are also famous for retro gaming enthusiasts and developers interested in creating software for older gaming consoles. Emulators for platforms like the Super Nintendo, PlayStation, and Game Boy allow users to play classic games and experiment with game development without owning the original hardware.

However, emulators have limitations. While they provide an effective means of testing, they may not always replicate the behavior of real devices accurately. Emulators might lack support for certain hardware features or struggle to mimic specific device characteristics faithfully. Furthermore, emulators may consume significant system resources, potentially impacting the performance of the development machine. The decision to use a development device or an emulator depends on various factors, including the nature of the project, budget constraints, and the availability of physical hardware.

For projects that involve web development or software targeting desktop operating systems, a development device in the form of a personal computer or laptop is the most practical choice. Developers can set up a robust development environment and test software in a native desktop environment.

The choice between a development device and an emulator in mobile app development depends on factors such as budget, target audience, and the need for real-world testing. Developing for iOS typically requires access to physical Apple devices for accurate testing, while Android developers can leverage both emulators and physical devices. Emulators are valuable for quickly iterating and debugging code, but physical devices are essential for ensuring a seamless user experience.

IoT development often necessitates the use of physical development devices, as they directly interact with specialized hardware components and sensors. Developers can connect to these devices via USB, Wi-Fi, or other communication protocols, enabling them to effectively develop, test, and debug IoT applications.

Choosing between a development device and an emulator hinges on project requirements and objectives. Development devices offer a real-world testing ground and are indispensable for certain types of development. On the other hand, Emulators provide a cost-effective and efficient means of testing, particularly for mobile app development and platform emulation. Ultimately, the decision should align with the project's specific needs, ensuring that software is thoroughly tested and optimized for its target platform.

CHAPTER III

The Basics of Programming

Introduction to programming languages for mobile app development (Java, Kotlin, Swift, etc.)

Mobile app development has become a cornerstone of our digital age, with billions of smartphone users relying on apps for everyday tasks and entertainment. Behind these apps' sleek and intuitive interfaces lies a complex web of code, and choosing the right programming language to build them is a pivotal decision. In this section, we'll explore some of the primary programming languages used in mobile app development, including Java, Kotlin, Swift, and more, shedding light on their strengths and applications in the ever-evolving landscape of mobile software development.

Java is one of the most enduring and ubiquitous programming languages in mobile app development. It is the primary language for building Android applications, powering millions of smartphones and tablets globally. Java's widespread adoption is partly due to its platform independence, making it highly portable across various devices and operating systems.

Java's robustness, extensive libraries, and developer-friendly features have made it a staple in Android development. It excels in building performance-oriented apps that handle complex tasks efficiently. Its object-oriented nature promotes modular and maintainable code, a crucial aspect in large-scale app development. While Java remains a powerful choice, it's worth noting that it has been supplemented by newer languages, particularly Kotlin, which offers a more concise and expressive alternative.

Kotlin burst onto the Android development scene in 2017, quickly gaining traction as an official alternative to Java for Android app development. JetBrains developed Kotlin to address some of Java's limitations while maintaining seamless compatibility with existing Java codebases. It introduced modern programming features like null safety, concise syntax, and enhanced expressiveness.

Kotlin's appeal lies in its ability to streamline Android app development. Its concise syntax reduces boilerplate code, leading to faster development cycles and fewer potential bugs. Kotlin's null safety features also help developers avoid null pointer exceptions, a common source of crashes in Android apps. The language's interoperability with Java allows developers to gradually transition from Java to Kotlin, making it an attractive choice for both new and existing Android projects.

Swift reigns supreme on the Apple side of the mobile app development landscape. Developed by Apple, Swift was introduced in 2014 as a replacement for Objective-C, the previous iOS and macOS app development standard. Swift's clean syntax, performance optimization, and emphasis on safety and readability have endeared it to Apple developers worldwide.

Swift delivers high-performance iOS, macOS, watchOS, and tvOS applications. Its syntax is designed to be approachable for newcomers to programming while remaining powerful enough to handle the most complex app development tasks. Swift's memory management model leverages automatic reference counting (ARC) to optimize memory usage and prevent common programming errors, making it a robust choice for building apps that run seamlessly across Apple devices.

Cross-platform app development has gained momentum in recent years, driven by the desire to write code once and deploy it on multiple platforms. With its versatility and widespread use, JavaScript is the programming language of choice for many cross-platform development frameworks. These frameworks, such as React Native, Flutter, and Xamarin, allow developers to create apps for both iOS and Android using a single codebase.

In combination with web technologies like HTML5 and CSS, JavaScript offers developers a familiar and accessible path to cross-platform app development. It leverages a "write once, run anywhere" philosophy, reducing development time and resources. JavaScript-based frameworks enable developers to access platform-specific features and deliver native-like experiences to users.

While the cross-platform approach offers efficiency and cost savings, it may involve trade-offs in terms of performance and access to platform-specific functionalities. When choosing this route, developers must carefully evaluate project requirements and user experience expectations.

The world of mobile app development is a diverse and dynamic one, with a wide array of programming languages catering to different platforms and development goals. Java and Kotlin dominate the Android ecosystem, offering performance and flexibility. On the other hand, Swift stands as the Apple ecosystem's linchpin, powering iOS, macOS, and more with its performance and safety features.

Cross-platform development with JavaScript and HTML5 has emerged as an efficient and cost-effective option, allowing developers to create apps for several different platforms using a single codebase. This approach emphasizes speed of development and broad reach but may require compromises regarding platform-specific optimizations.

Ultimately, the choice of programming language for mobile app development depends on project objectives, target platforms, and developer preferences. The evolving nature of technology continually introduces new languages and frameworks, expanding the possibilities for mobile app development. As developers navigate this multilingual landscape, they are equipped with an ever-expanding toolbox to bring innovative and impactful mobile applications to life.

Variables, data types, and basic syntax

In the realm of programming, variables, data types, and basic syntax serve as the foundational elements upon which every software application is constructed. These fundamental concepts empower developers to create code to process information, make decisions, and perform tasks. In this section, we delve into the essential concepts of variables, data types, and basic syntax, illuminating their significance in the programming world.

At the core of programming lies the concept of variables. A variable is like a container holding different data types, such as numbers, text, or complex structures. Variables are the building blocks for storing and manipulating information within a program. When a variable is declared, the programming language reserves a specific memory location to store the data associated with that variable.

Variables are essential for dynamic programming because they allow the program to work with data that can change during execution. For example, in a weather app, a variable could store the current temperature, which might fluctuate throughout the day. By using variables, the app can adapt to changing conditions and provide accurate information to the user.

Data types are crucial in programming by specifying the kind of data that can be stored in a variable. Each programming language offers a set of predefined data types, each with its own characteristics and limitations. Common data types include integers (whole numbers), floating-point numbers (numbers with decimal points), strings (text), and booleans (true or false values). Choosing

the right data type is essential because it determines how the data is stored in memory and how operations are performed on it. For example, using an integer data type for a variable that should hold a person's age makes sense, but using it for a variable that should store a person's height in meters would be inappropriate due to the need for decimal precision.

Modern programming languages often allow developers to define custom data types representing more complex structures and objects. These custom data types enhance code readability and maintainability by encapsulating related pieces of data and functionality into a single unit.

The basic syntax of a programming language defines the rules for writing code. It dictates how instructions should be structured, how statements should be terminated, and how code blocks are delimited. Syntax errors occur when the code violates these rules, making it essential for programmers to adhere to the language's syntax guidelines.

For example, in many programming languages, statements are terminated with a semicolon (;), and code blocks are enclosed within curly braces ({ }). Missing or misplaced semicolons or curly braces can lead to syntax errors, preventing the program from running correctly.

Furthermore, basic syntax includes rules for naming variables, functions, and other program elements. These rules often involve restrictions on characters and require variable names to be descriptive and meaningful. Good naming practices enhance code readability and make it easier for developers to comprehend the purpose of each element.

In the world of programming, variables, data types, and basic syntax form the language through which software applications communicate with computers. Variables act as data containers, allowing programs to work with dynamic information. Data types define the nature of the data, ensuring that it is stored and processed correctly. Basic syntax establishes the rules and structure of code, ensuring that instructions are correctly constructed. Aspiring programmers often start by mastering these fundamental concepts, and experienced developers rely on them daily. By understanding how variables hold data, how data types define data characteristics, and how syntax governs code structure, programmers gain the ability to craft logic and functionality that powers a wide range of software applications, from simple scripts to complex systems. Ultimately, these building blocks are the essence of programming, forming the bridge between human logic and machine execution, enabling the creation of software solutions that shape our digital world.

Understanding control structures (if statements, loops)

Control structures are the foundation of programming, providing the means to make decisions, repeat tasks, and create logic in code. Two fundamental control structures, if statements and loops, play a pivotal role in directing the flow of a program. This section explores the significance of these control structures, their syntax, and how they empower developers to create dynamic and responsive software.

If statements, also known as conditional statements, are essential for decision-making in programming. They allow a program to execute different code blocks based on specific conditions. At its core, an if statement evaluates a condition as either true or false and performs an action accordingly.

An if statement's basic syntax is the keyword "if" which is followed by a condition that is contained in parenthesis. If the condition evaluates to true, the code block following the if statement is executed. The code block is skipped if the condition assesses to false. For more complex decision-making, if statements can be extended with "else if" and "else" clauses to handle multiple conditions and fallback actions.

If statements find applications in a wide range of scenarios. For instance, in a weather app, an if statement might check if the current temperature is above a certain threshold and display a heat advisory message if true. In a game, an if statement can determine if a player's score qualifies for a high score and prompt them to enter their name.

Loops are control structures that enable the repetition of code, making them invaluable for tasks that involve performing the same operation multiple times. Loops help streamline repetitive tasks, enhance code efficiency, and enable the processing of large datasets.

"for" loops and "while" loops are the two primary categories of loops. A "for" loop specifies a range and iterates through it a fixed number of times. In contrast, a "while" loop repeats as long as a specified condition remains true.

In a "for" loop, the syntax includes defining a loop variable, setting the loop's condition, and specifying how the loop variable changes with each iteration. For example, a "for" loop can be used to iterate through an array of numbers and calculate their sum.

On the other hand, a "while" loop relies solely on a condition that is checked before each iteration. The loop continues to perform as long as the condition remains true. For instance, a "while" loop can be used to prompt a user for input until a valid response is received.

Loops are invaluable for automating tasks like data processing, generating sequences, and searching through arrays. They reduce code duplication, enhance readability, and improve code maintainability. However, programmers must be cautious when working with loops to avoid issues like infinite loops, which can lock up a program.

One of the strengths of programming lies in the ability to combine control structures to create intricate logic. For instance, an if statement can be nested inside a loop, allowing developers to make decisions based on each iteration's results. This combination of control structures enables the development of sophisticated algorithms and complex behaviors in software.

Imagine a situation in which a developer wants to create a program that simulates a weather forecasting system. They might use a "for" loop to iterate through a range of future dates, and within that loop, they could use an if statement to check weather conditions for each date and provide corresponding forecasts.

Control structures, including if statements and loops, are the heart of program flow in software development. They enable developers to create dynamic, responsive, and efficient applications by making decisions and automating repetitive tasks. Understanding how to use these control structures and their syntax and nuances is a fundamental skill for programmers. Developers can create logic that converts static code into dynamic software solutions that support a variety of scenarios and user interactions by utilizing the tools that they have at their disposal. Control structures are the building blocks that empower developers to breathe life into their code and bring software to life.

CHAPTER IV

Creating Your First Mobile App

Building a "Hello World" app

Creating a "Hello World" app is a rite of passage for every aspiring programmer, marking the initial journey into software development. This simple yet essential exercise serves as an introduction to programming languages, development environments, and the process of turning code into a functional application. In this section, we explore the significance of the "Hello World" app, the steps involved in its creation, and the broader implications for those embarking on their programming journey.

The "Hello World" app represents the very essence of a programmer's craft. It is a minimalistic program designed to accomplish a straightforward task: displaying the words "Hello, World!" on the screen. While this may seem trivial, the implications are profound. The "Hello World" app introduces novice programmers to the core concepts of a programming language, including syntax, variables, and output.

The primary objectives of creating a "Hello World" app are twofold. First, it offers a tangible entry point into the world of coding, enabling beginners to see their code in action and gain a sense of accomplishment. Second, it provides a foundation upon which more complex applications can be built. The skills learned in creating a

"Hello World" app are building blocks for future programming endeavors.

The method used to create a "Hello World" app can change based on the development environment and programming language selected. However, the core steps remain consistent:

Selecting a Programming Language: Choose a programming language in which you want to write your "Hello World" app. Common choices include Python, JavaScript, Java, C++, and many more. The choice of language may depend on personal preference, project requirements, or the desire to learn a specific language.

Setting Up the Development Environment: Install the tools and software required for programming in your chosen language. This often includes an Integrated Development Environment (IDE) or a code editor. For writing, testing, and debugging code, IDEs such as PyCharm, Xcode, or Visual Studio Code offer a comprehensive interface.

Writing the Code: Open your code editor or IDE and create a new project or file. Write the code necessary to display the "Hello, World!" message. The specific code syntax will vary depending on the chosen programming language. For example, in Python, the code might be as simple as print("Hello, World!"), while in Java, it might involve creating a class and a primary method.

Compiling and Running: Depending on the programming language, you may need to collect the code into an executable form. Some languages, like Python and JavaScript, are interpreted, meaning you can run the code directly. You'll need to compile the code in compiled languages like C++ or Java first. Once compiled, execute the program to see the "Hello, World!" message displayed.

Observing the Output: After running the program, watch the output, which should display the "Hello, World!" message on the screen. This moment signifies the successful execution of your first program.

Reflecting and Learning: Take a moment to reflect on the experience. Consider the code you've written, the syntax you've used, and the process of turning code into a functioning application. This reflection is an essential part of the learning process.

While a "Hello World" app is a simple exercise, it carries broader implications for those pursuing a career in software development. Beyond its educational value, it symbolizes the transition from a beginner programmer to someone who can write, execute, and understand code.

The "Hello World" app is also a testament to the universality of programming languages. Regardless of the language chosen, the core concepts of coding remain consistent. Once you've mastered the basics of one language, you'll find it easier to learn and adapt to others, opening up a world of opportunities in various fields of software development.

Furthermore, the "Hello World" app serves as a reminder of the power of coding. With a few lines of code, you can create software that performs a specific task or solves a particular problem. This ability to transform ideas into functional applications is at the heart of software development, and it's a skill that continues to be in high demand in today's digital age.

In conclusion, building a "Hello World" app is not just a programming exercise; it's a symbolic initiation into software development. It embodies the essence of coding, introduces core programming concepts, and marks the beginning of a journey into a field that is as diverse as it is dynamic. As you create your "Hello World" app, remember that it's not just about displaying a simple message; it's about embarking on a path of discovery, learning, and endless possibilities in the world of code.

Exploring the app's structure (UI, code)

In the world of mobile app development, understanding an application's structure is like deciphering a building's blueprint. It allows developers to comprehend how different elements interact, and it's crucial for building, maintaining, and troubleshooting apps effectively. The app's structure typically consists of two key components: the User Interface (UI) and the underlying code. In this section, we'll delve into the significance of exploring these two facets of an app's structure and how they collectively shape the user experience.

An app's User Interface (UI) represents the visual and interactive elements that users interact with on their devices. It encompasses everything from the layout and design of screens to the placement of buttons and the overall user experience. A well-designed UI is critical for attracting and retaining users, as it directly impacts how they perceive and interact with the app.

The UI typically comprises multiple screens or views, each with its own purpose and functionality. For instance, a social media app might have screens for news feeds, user profiles, and messaging. Each screen is designed to present specific information and enable particular actions, tailored to the app's functionality.

User Interface elements include widgets, such as buttons, text fields, images, and navigation bars, that users interact with to perform actions or access information. Additionally, the UI encompasses the app's layout, which defines the arrangement of these elements on the screen. Layout considerations include screen orientation (portrait or landscape), responsiveness to different device sizes, and adherence to design guidelines for a cohesive and visually pleasing experience.

Exploring the UI involves evaluating how these elements are organized, styled, and interacted with. UI design tools and frameworks are crucial in creating visually appealing and user-friendly interfaces. Understanding the UI structure allows developers to make informed decisions about layout adjustments, color schemes, and user interactions, ensuring a seamless and enjoyable user experience.

Beneath the polished exterior of the UI lies the underlying code, which is responsible for driving the app's functionality. This code comprises algorithms, data structures, and logic that dictate how the app processes data, responds to user input, and communicates with external services or databases.

The structure of the code depends on the chosen programming language and the app's architecture. For example, mobile apps are commonly developed using languages like Java, Swift, or Kotlin, each with its own coding conventions and best practices. The code is organized into classes, functions, and modules, with each component responsible for specific tasks and functionalities.

Exploring the code involves examining how the app stores, retrieves, and manipulates data. This may include interactions with databases to store user information or retrieve content, processing user input to trigger actions, and implementing algorithms to support various features, such as search, recommendations, or notifications.

The code structure also encompasses the app's architecture, defining how different components interact. Common architectural patterns include Model-View-Controller (MVC), Model-View-ViewModel (MVVM), and Clean Architecture. The choice of architecture influences code organization, maintainability, and scalability, making it a crucial consideration for app development.

Understanding the code structure is essential for developers to maintain and extend the app's functionality effectively. It allows for easier debugging, optimization, and implementing new features. Code readability and maintainability are paramount, as apps often undergo updates and enhancements throughout their lifecycle. A well-structured codebase is a foundation for long-term app sustainability.

In the app development world, the UI and code are intertwined, each influencing the other to create a cohesive and user-centric experience. A well-designed UI can enhance an app's appeal, but the underlying code empowers the app to deliver on its promises.

Successful app development involves a collaborative effort between designers and developers to align the UI and code with the app's objectives and user needs. Designers focus on aesthetics, usability, and user flow, while developers implement the functionality that brings the app to life.

As developers explore the app's structure, they bridge the gap between the UI's visual elements and the code's functional components. This understanding enables them to optimize performance, troubleshoot issues, and continuously refine the user experience. It's a synergy that underlies every successful app, from simple "Hello World" applications to complex, feature-rich software.

In conclusion, exploring the structure of a mobile app, encompassing both the User Interface (UI) and underlying code, is a fundamental aspect of app development. The UI shapes the user's initial impression and interaction with the app, while the code dictates how the app functions and delivers on its promises. By understanding and refining both aspects, developers can craft user-centric experiences that meet the expectations of today's discerning app users. The synergy between UI and code is the key to creating successful and impactful mobile applications that resonate with audiences worldwide.

Running your app on an emulator or device

After the meticulous process of designing and coding your mobile app, the next crucial step is to see it in action. To do this, you have two primary options: running your app on an emulator or testing it on a physical device. Both methods serve essential purposes in the development lifecycle and offer unique advantages. In this section, we'll explore the significance of running your app on an emulator or device, the steps involved, and how these approaches contribute to the refinement and success of your mobile application.

Emulators are software-based tools that replicate the behavior of a specific mobile device or platform on your development machine. They allow you to test your app without physical hardware, making them a valuable resource for development. Emulators are available for various platforms, including Android, iOS, and even specialized devices like gaming consoles.

One of the primary advantages of emulators is their versatility. Developers can use emulators to test their apps on different device configurations, screen sizes, and operating system versions, ensuring compatibility across various scenarios. Emulators also support features like screen recording and the ability to simulate various network conditions, making them valuable for debugging and user experience testing.

Emulators are particularly useful for early-stage development and debugging. They provide a rapid and efficient way to iterate on your code, allowing you to make changes and observe their impact quickly. Emulators also enable you to test your app under various conditions, such as device rotations, screen resolutions, and simulated user interactions.

While emulators offer convenience and versatility, physical devices provide the closest approximation to real-world user experiences. Testing your app on actual smartphones and tablets allows you to assess how it performs in the hands of users. Physical devices can reveal issues that may not be apparent in an emulator, such as hardware-specific bugs, touch screen responsiveness, or device-specific behaviors. Physical devices are essential for fine-tuning the user experience and ensuring that your app functions as expected on the devices your target audience uses. For example, if you're developing an Android app, testing it on various Android devices helps identify compatibility issues and ensures a consistent experience across different manufacturers and models.

Another advantage of physical devices is the ability to evaluate your app's behavior in real-world network conditions. Mobile networks can introduce latency, variable bandwidth, and connectivity issues that may impact your app's performance. Testing on a physical device connected to a cellular network allows you to assess how your app handles these challenges and optimize it accordingly.

Regardless of whether you choose emulators or physical devices, the process of running your app follows a general sequence:

Configuration: Ensure your development environment is set up correctly for the chosen platform. This includes installing the necessary SDKs (Software Development Kits) and tools, configuring development options, and connecting physical devices if applicable.

Build and Deployment: Compile your app's code into an executable form suitable for running on the target platform. This may involve creating an APK (Android Package) for Android apps or an IPA (iOS App Store Package) for iOS apps. Deploy the app to the emulator or physical device.

Testing and Debugging: Launch the app on the emulator or device, and interact with it to test various features and functionalities. Use debugging tools and logs to identify and resolve any issues that may occur during testing.

User Experience Evaluation: On physical devices, pay close attention to the user interface, responsiveness, and overall user experience. Test different scenarios and use cases to ensure that your app performs reliably.

Performance Optimization: If you encounter performance issues or bugs, make necessary code adjustments and repeat the testing process. Continuously refine and optimize your app based on testing results.

User Acceptance Testing (UAT): If applicable, involve beta testers or end-users in the testing process. Collect feedback, identify potential issues, and make further improvements based on user input.

Submission: Once you're satisfied with your app's performance and user experience, prepare it for submission to app stores including Google Play or the Apple App Store. Follow the respective guidelines and requirements for app submission.

Emulators and physical devices play complementary roles in developing and testing mobile apps. Emulators offer convenience and versatility for rapid development and early-stage testing. They allow developers to iterate quickly and test under various conditions.

On the other hand, physical devices provide the closest approximation to real-world user experiences. They are essential for fine-tuning the app, ensuring compatibility, and evaluating performance under actual network conditions.

A balanced approach combining emulators and physical devices often yields the best results. Developers can use emulators for initial development and debugging and then transition to physical devices for user experience evaluation and real-world testing.

Running your app on an emulator or device marks a significant milestone in the development journey. It's when your creation transitions from lines of code to a tangible, interactive experience. Both emulators and physical devices serve essential roles in this process, offering unique advantages and perspectives. Emulators provide convenience, versatility, and rapid testing capabilities, making them ideal for early development and debugging. Physical devices offer real-world insights, enabling you to fine-tune the user experience and ensure compatibility with various device configurations.

Choosing between emulators and physical devices depends on your development goals and needs. Combining both approaches can provide a comprehensive and effective testing strategy, ensuring your app is ready to meet users' expectations in the ever-evolving world of mobile applications.

CHAPTER V

User Interface Design

Principles of good UI/UX design

In the digital era, user interface (UI) and also user experience (UX) design are the cornerstones of creating engaging and successful applications. Whether it's a mobile app, website, or software program, the design of the user interface and the overall user experience significantly influence how users interact with and perceive a product. To excel in UI/UX design is to understand and apply a set of fundamental principles that guide the creation of aesthetically pleasing, functional, and user-centric interfaces. This section explores these principles and their pivotal role in crafting memorable digital experiences.

At the heart of UI/UX design is the principle of user-centered design. This principle emphasizes understanding the target audience's needs, preferences, and behaviors. To develop an interface that resonates with users, designers must conduct user research, gather feedback, and empathize with the end-users. By putting the user at the center of the design process, designers can develop interfaces that are intuitive, efficient, and tailored to the user's context and goals.

Consistency is a cornerstone of good UI/UX design. It involves maintaining a uniform design language and behavior throughout the application. Consistency ensures that users can predict how different elements will behave and where to find specific features. This principle extends to visual elements, interaction patterns, terminology, and branding. A consistent design fosters a sense of familiarity and comfort for users, making navigating and using the application easier.

The principle of simplicity advocates for minimalism in design. It encourages designers to eliminate unnecessary complexity and distractions, focusing on delivering a straightforward and streamlined user experience. A simple design reduces cognitive load, making it easier for users to understand the interface, complete tasks, and achieve their goals. Cluttered interfaces with excessive features and information can overwhelm users and lead to frustration.

Visual hierarchy is the practice of using design elements to guide the user's attention and prioritize information. It involves arranging elements such as text, images, and buttons to communicate their relative importance. Elements that are more critical to the user's task should be more prominent and easily accessible. Visual hierarchy ensures that users can quickly identify what's essential and easily navigate the interface.

Feedback and responsiveness are vital aspects of UI/UX design. Users expect immediate feedback when interacting with an application. Whether it's pressing a button, filling out a form, or swiping on a touchscreen, the interface should respond promptly to user actions. Providing feedback through animations, notifications, and changes in element appearance assures users that their actions are acknowledged, reducing uncertainty and frustration.

Accessibility is a principle that promotes the design of interfaces that people of all abilities can use. Designers should consider factors such as screen readers for the visually impaired, keyboard navigation for those with motor disabilities, and clear and legible text for all users. Creating accessible interfaces ensures inclusivity and aligns with legal requirements and ethical design practices.

UI/UX design is an iterative process that involves continuous improvement. Designers should embrace testing and feedback as essential components of the design lifecycle. Usability testing, A/B testing, and user feedback sessions help identify areas for improvement and validate design decisions. Iteration allows designers to refine the user experience, fix issues, and adapt to changing user needs and preferences.

The goal of good UI/UX design is to create a meaningful and pleasurable user experience, not just an attractive one. These principles serve as a compass, guiding designers toward crafting interfaces that resonate with users, make their tasks easier, and leave a lasting positive impression. By putting users at the center of the design process, maintaining consistency, embracing simplicity, and adhering to principles of accessibility and feedback, designers can create digital experiences that stand out in a crowded digital landscape. Ultimately, UI/UX design is an art that combines creativity, empathy, and user-centric thinking to transform software and applications into tools that users love and rely on.

Designing app layouts and elements

The design of app layouts and elements is critical to creating user-friendly and visually appealing applications. A well-designed layout influences how users interact with an app and plays a significant role in user retention and satisfaction. From the placement of buttons to the choice of colors and typography, every design decision impacts the overall user experience. This section delves into the principles and best practices of designing app layouts and elements that enhance usability and resonate with users.

The layout of an app refers to the arrangement and organization of its visual elements, such as buttons, text, images, and navigation menus, within the screen space. Effective layout design is rooted in user-centered principles and aims to optimize the user experience by ensuring that content is accessible, engaging, and easy to navigate.

One of the fundamental goals of app layout design is to guide users' attention and actions. A well-structured layout directs users' focus toward essential elements and facilitates the completion of tasks. It considers the flow of information and user interactions, aligning them with the app's objectives and user needs.

Hierarchy and visual flow are fundamental principles in layout design. Hierarchy involves arranging elements in a way that communicates their relative importance. Elements that are more critical to the user's task should be more prominent, while secondary elements should be less obtrusive. On the other hand, visual flow guides users' eyes through the content in a natural and intuitive manner.

To create a sense of hierarchy, designers use size, color, typography, and placement techniques. For example, a prominent call-to-action button with a contrasting color and a larger font size draws attention and invites user interaction. Visual flow is established by aligning elements in a logical sequence, ensuring that users can follow the content from left to right and from top to bottom without confusion.

Consistency is a cornerstone of effective layout design. Consistent layouts establish familiarity and predictability for users, making navigating and interacting with an app easier. Designers maintain consistency in visual elements, such as buttons, icons, and navigation bars, as well as in the arrangement of content across screens.

Branding is crucial in layout design, as it conveys the app's identity and reinforces its recognition. Design elements like logos, color schemes, and typography should align with the app's branding guidelines to create a cohesive and memorable user experience.

Adaptability and responsiveness are paramount in layout design in today's multi-device landscape. Users access apps on various screen sizes and orientations, ranging from smartphones and tablets to desktops and wearables. Designing responsive layouts ensures the app adapts seamlessly to different devices and screen resolutions, providing a consistent user experience.

Responsive design utilizes flexible grids and media queries to adjust the layout and content presentation based on the device's characteristics. It eliminates the need for separate designs for each device, streamlining development and maintenance.

User testing is a valuable phase in layout design. It entails gathering feedback from actual users to assess the layout's effectiveness and identify improvement areas. Usability testing, user interviews, and feedback surveys provide insights into the way users perceive and interact with the app.
Designers use user feedback to refine the layout, address usability issues, and make adjustments to enhance the user experience. Iteration is a continuous process that involves testing and refining the layout based on real-world user interactions and preferences.

Designing app layouts and elements is an art that combines aesthetics, psychology, and user-centered principles. A well-crafted layout guides users through an app, optimizes their interactions, and communicates the app's brand and identity. By applying principles of hierarchy, visual flow, consistency, adaptability, and responsiveness, designers can create layouts that resonate with users and enhance the overall user experience.

The evolution of technology and the diversity of devices demand adaptability and continuous improvement in layout design. User testing and iterative design are essential components of the process, ensuring layouts evolve to meet changing user needs and preferences. Ultimately, app success hinges on its ability to provide an intuitive and visually pleasing interface that captivates users and fosters engagement, making thoughtful layout design a fundamental aspect of app development.

Using layout tools and widgets

In the realm of app development, creating an intuitive and visually pleasing user interface (UI) is paramount to capturing and retaining users' attention. To achieve this, developers leverage layout tools and widgets. Layout tools help arrange UI elements, while widgets are pre- built UI components that enhance functionality and user engagement. In this section, we explore the significance of layout tools and widgets in app development, how they simplify the design process, and their role in crafting memorable user experiences.

Layout tools are instrumental in organizing and structuring the elements of an app's user interface. They provide developers with a visual canvas to arrange buttons, text, images, and other components, ensuring a cohesive and user-friendly layout. These tools eliminate the need for manual positioning and sizing of UI elements, simplifying the design process and enhancing efficiency.

One of Android's most commonly used layout tools is the XML-based layout system, which employs XML files to define the UI structure. Developers can specify UI elements' position, size, and alignment using XML attributes, resulting in a responsive and adaptable layout.

On the iOS platform, Interface Builder is a visual layout tool that allows developers to design app interfaces through a drag-and-drop interface, offering a similar level of convenience.

Layout tools also facilitate the creation of responsive designs that adjust to a wide range of screen sizes and orientations. This adaptability is crucial in today's multi-device landscape, where users access apps on smartphones, tablets, and desktops. These tools ensure a consistent user experience across devices by providing responsive layout options.

Widgets are pre-built UI components that extend an app's functionality and visual appeal. They serve as building blocks for constructing app interfaces and provide users with interactive elements that enhance engagement. Widgets encompass many UI elements, including buttons, text fields, sliders, progress bars, etc.

For instance, a button widget allows users to trigger actions, like submitting a form or navigating to another screen. A progress bar widget visually communicates the status of a task, while a slider widget enables users to adjust settings or values interactively. Widgets are versatile and can be customized to align with the app's design and functionality.

Beyond the standard UI elements, widgets can also include advanced components such as charts, graphs, and multimedia players. These widgets empower developers to create data-rich and visually captivating app interfaces. In e-commerce apps, for example, widgets displaying product images, reviews, and interactive shopping carts enhance the user experience and encourage user engagement.

The synergy between layout tools and widgets is pivotal in designing compelling app interfaces. Layout tools provide the canvas and structure for arranging widgets, allowing developers to position and align them precisely. Widgets, in turn, enrich the layout by introducing interactive and functional components that meet user needs and expectations.

Customizing widgets within the layout is a critical aspect of UI design. Developers can adjust attributes such as color, size, typography, and behavior to align with the app's branding and user interface guidelines. This level of customization ensures that the app's design remains cohesive and consistent, fostering brand recognition and enhancing the user experience.

Moreover, the use of widgets extends beyond static elements. They can be integrated with data sources, enabling dynamic content updates and real-time interactions. For instance, a weather app can utilize a widget to display live weather data sourced from a server, ensuring that users receive up-to-date information.

In the world of app development, the marriage of layout tools and widgets is the bedrock of crafting memorable and engaging user experiences. Layout tools provide the structure and adaptability needed to organize UI elements efficiently, ensuring a consistent and responsive design across devices. Conversely, widgets enrich the interface with interactive and functional components that enhance engagement and fulfill user needs.

Developers must balance the visual appeal and functionality of app interfaces, leveraging the capabilities of layout tools and widgets to create seamless and user-centric designs. By mastering and utilizing these tools effectively, developers can elevate their apps to stand out in the competitive landscape and leave a lasting impression on users. Ultimately, the harmonious interplay between layout tools and widgets is the key to building intuitive, engaging, and successful app interfaces that resonate with today's discerning users.

CHAPTER VI

Working with Data

Storing and retrieving data in mobile apps

Our everyday lives now wouldn't be the same without mobile apps, which easily handle everything from productivity and communication to shopping and entertainment. At the core of these apps lies a crucial component: the ability to store and retrieve data efficiently. Data management is the backbone of user interaction, whether it's user preferences, messages, photos, or transaction records. In this section, we explore the significance of storing and retrieving data in mobile apps, the various approaches and technologies involved, and their impact on the user experience.

Data management in mobile apps encompasses a broad spectrum of tasks, including storing user profiles, handling authentication, saving user-generated content, and managing application state. Storing and retrieving data efficiently is critical for providing a seamless and responsive user experience.

Consider a social media app. It must store user profiles, posts, comments, and media files while ensuring fast retrieval when users scroll through their feeds. An e-commerce app must manage product listings, user accounts, shopping carts, and order histories. In both cases, effective data management directly influences user engagement and satisfaction.

Mobile apps typically employ local and remote storage solutions to manage data effectively. Local storage refers to storing data on the user's device, such as using a device's file system or a local database. Local storage is ideal for storing small amounts of data, user preferences, and frequently accessed content.

Remote storage, on the other hand, involves storing data on remote servers or cloud platforms. Remote storage is essential for managing large datasets, facilitating data synchronization across multiple devices, and ensuring data security and backup. Apps that rely on user accounts, real-time updates, or collaboration among users often use remote storage.

For local storage, mobile app developers have several options at their disposal. SQLite, a lightweight and embedded relational database, is popular for storing structured data. It allows developers to create and manage tables, execute SQL queries, and ensure data integrity. SQLite is particularly suitable for apps that require complex data querying and filtering.

Another local storage option is SharedPreferences for Android and UserDefaults for iOS, which provide a simple key-value store for storing user preferences and small amounts of data. For unstructured data, apps can also utilize file storage systems to save images, documents, and other media files locally.

Remote storage in mobile apps often involves connecting to server-side APIs (Application Programming Interfaces) or utilizing cloud-based services. RESTful APIs, which use HTTP methods for data retrieval and manipulation, are commonly used for data exchange between mobile apps and servers. These APIs provide a standardized and straightforward way to interact with remote data sources.

Cloud-based storage services, like Amazon S3, Google Cloud Storage, or Firebase Realtime Database, offer scalable and reliable solutions for remote data storage. These services handle data synchronization, security, and scalability, allowing app developers to focus on app functionality rather than infrastructure management.

Data security and privacy are paramount considerations in data management for mobile apps. App developers must implement encryption techniques to protect sensitive user data, especially when transmitting data over networks. Secure authentication mechanisms, such as OAuth or token-based authentication, ensure only authorized users can access data.

Moreover, compliance with data protection regulations, like GDPR (also known as General Data Protection Regulation) in Europe or the CCPA (also known as California Consumer Privacy Act) in the United States, is crucial. Mobile app developers must be vigilant in handling user data, providing transparency in data collection and usage, and enabling users to exercise control over their data.

In the world of mobile apps, effective data management is the linchpin that holds together seamless user experiences. It enables apps to load content quickly, offer personalized experiences, and ensure data security and privacy. The capacity to store and retrieve data efficiently transforms an app from a static piece of software into a dynamic and user-centric tool.

Whether it's a messaging app, a fitness tracker, or a productivity tool, the success of a mobile app often hinges on how well it manages and utilizes data. By adopting the right storage solutions, employing secure practices, and adhering to data privacy regulations, app developers can build user experiences that engage, delight, and foster user trust. In the ever-evolving landscape of mobile apps, data management remains at the forefront of innovation and user satisfaction, making it an indispensable component of app development.

Introduction to databases

In today's data-driven world, the role of databases is paramount. Databases serve as the backbone of information management for countless organizations, from small businesses to multinational corporations, and even in our everyday lives when we interact with various applications and websites. A database is a structured collection of data that enables for efficient storage, retrieval, and manipulation of information. This section will comprehensively introduce databases, exploring their importance, basic concepts, types, and the key technologies that power them.

Databases play a pivotal role in the modern digital landscape, enabling the storage and management of vast amounts of data. The importance of databases lies in their ability to organize and structure information to facilitate easy access and retrieval. Without databases, data would be unorganized, scattered, and difficult to manage, hindering decision-making processes, business operations, and virtually every aspect of our lives that relies on information.

To understand databases better, it's essential to grasp some fundamental concepts. The primary building block of a database is the data itself, which can be any piece of information, such as names, numbers, or even multimedia content like images and videos. Data is organized into tables, each representing a specific entity or concept. These tables consist of rows and columns, with each row illustrating a unique record, and each column representing an attribute or field of the records.

One of the key principles of databases is the concept of data integrity. Data integrity ensures that the information stored in the database is accurate, consistent, and reliable. To maintain data integrity, databases often use constraints and rules to enforce data validity. For example, a database may require that all email addresses follow a specific format or that certain fields cannot be left empty.

Databases also provide the ability to establish relationships between different tables. This is done through keys, which are unique identifiers for records in a table. Common types of keys include primary keys, which uniquely identify each record in a table, and foreign keys, which establish connections between tables. These relationships are crucial for modeling complex data structures and efficiently retrieving related information. There are several types of databases, each designed for specific use cases. The two primary categories are relational databases and non-relational databases. Relational databases, such as MySQL, PostgreSQL, and Oracle, are structured around tables with predefined schemas. They are excellent for managing structured data and are widely used in applications where data consistency and reliability are critical, such as financial systems and customer relationship management (CRM) software.

On the other hand, non-relational databases, also known as NoSQL databases, are more flexible in terms of data structure. They are well-suited for handling unstructured or semi-structured data, making them popular for web applications, big data analytics, and content management systems. Examples of NoSQL databases include MongoDB, Cassandra, and Redis.

The selection between relational and non-relational databases depends on the specific needs of an application. Relational databases excel in scenarios where data relationships are well-defined and must be maintained rigorously. In contrast, NoSQL databases offer greater scalability and flexibility, making them a better fit for applications that require handling large volumes of rapidly changing data.

Developers and users rely on a database management system (DBMS) to interact with databases. A DBMS is software that provides an interface for creating, modifying, and querying databases. It acts as an intermediary between the user and the physical database, handling data storage, retrieval, and security tasks. Popular DBMSs include MySQL, Microsoft SQL Server, and MongoDB.

SQL, also called as Structured Query Language, is a standardized language utilized to communicate with relational databases. It allows users to define, manipulate, and query data in a database. SQL statements can perform various operations, such as selecting data from tables, inserting new records, updating existing records, and deleting records. The power of SQL lies in its ability to perform complex operations with a relatively simple and human-readable syntax.

Beyond SQL, there are also programming languages and frameworks that facilitate database interaction. For example, Python provides libraries like SQLAlchemy for working with databases, while web development frameworks like Ruby on Rails and Django offer built-in support for database integration, simplifying the process of creating web applications that interact with databases.

Security is a crucial aspect of database management. Databases often contain sensitive information, such as personal data, financial records, and intellectual property. As a result, they are prime targets for cyberattacks. DBMSs employ various security mechanisms to protect data, including authentication, authorization, encryption, and auditing. Organizations must implement robust security practices to safeguard their databases and their information.

In conclusion, databases are the cornerstone of modern information management. They provide the means to store, retrieve, and manipulate data efficiently, making them indispensable in today's data-driven world. Understanding the fundamental concepts of databases, including data organization, integrity, and relationships, is essential for anyone working with data. Moreover, the choice between relational and non-relational databases, coupled with the use of database management systems and query languages like SQL, empowers organizations and individuals to harness the full potential of their data while ensuring its security. As we continue to generate and rely on ever-increasing amounts of data, the role of databases will only become more significant in shaping our digital landscape and driving innovation across various industries.

Implementing data storage in your app

Mobile app development is a thrilling journey filled with innovation and creativity. Yet, it's also a journey riddled with challenges, one of the most formidable being implementing data storage. As developers strive to bring their vision to life, they often encounter a myriad of choices, challenges, and best practices when storing and managing data efficiently. This section will explore the art of implementing data storage in your mobile app, from the choices available to the best practices that empower developers to create apps that handle data effectively, securely, and seamlessly.

Before diving into the technical aspects of data storage, it's essential to understand your app's data storage requirements. Consider the data types your app will handle, such as text, images, videos, or structured information like user profiles or product listings. Determine the volume of data your app will manage and the expected growth over time. Consider data access patterns, including read and write operations, data retrieval, and data modification. A clear understanding of these requirements will guide your decisions throughout development.

Mobile app developers have several options when it comes to data storage. Local storage, including SQLite databases for structured data and SharedPreferences (on Android) or UserDefaults (on iOS) for key-value pairs, is ideal for small to moderate-sized datasets that must be readily available offline. File storage, supported on Android and iOS, is suitable for managing larger files, such as images, audio, or video. For more advanced use cases, cloud-based solutions like Amazon S3, Google Cloud Storage, or Firebase Cloud Storage provide scalability, real-time updates, and data synchronization across multiple devices.

The choice of data storage solution should align with your app's specific needs, performance requirements, and scalability considerations. Careful consideration at this stage can save time and effort in the long run.

Data security and privacy are paramount in mobile app development. When implementing data storage, it's crucial to prioritize security measures. Utilize encryption mechanisms to protect data both at rest and in transit. Implement robust authentication to ensure that only authorized users can access specific data, and use role-based access control (RBAC) to define and enforce data access permissions. Additionally, follow the principle of data minimization, collecting and storing only the data necessary for the app's functionality. Handling sensitive or personally identifiable information (PII) with the utmost care is essential to maintain user trust and compliance with data protection regulations.

Offline data access is a crucial aspect of mobile app development. Users expect seamless app functionality even when an internet connection is unavailable. To support offline access, consider implementing data caching and synchronization mechanisms. Use local storage or database solutions to store a subset of data locally, allowing users to interact with the app's content without an internet connection. Implement background data synchronization processes that update local data when a network connection is available. This ensures users can continue using your app, even in challenging network conditions.

Data loss can be catastrophic for users and result in negative app reviews. Implement regular data backup and recovery mechanisms to safeguard user data. Many cloud providers offer automated backup and restore features for apps using cloud-based storage. Consider providing users with backup options for local data storage, such as exporting data to cloud storage services or enabling local backups. By offering these features, you protect user data and enhance user trust and satisfaction.

Mobile apps are continually evolving, so data structures and schemas may change over time. When updating your app, plan for data migration and versioning to ensure a seamless transition for existing users. Implement robust migration scripts and versioning mechanisms to handle changes to the data schema without causing data loss or data incompatibility issues. This proactive approach minimizes disruptions and ensures that users can smoothly transition to updated versions of your app.

Implementing data storage in your mobile app is critical to app development. By understanding your app's data storage requirements, choosing the right storage solution, prioritizing data security and privacy, supporting offline data access, implementing data backup and recovery mechanisms, and planning for data migration and versioning, you can ensure that your app effectively manages and utilizes data to provide a seamless and reliable user experience. Data storage is not merely a technical aspect of app development; it's a fundamental element that underpins user engagement and satisfaction. Through thoughtful and strategic data storage implementation, you can create apps that users trust, rely on, and enjoy.

CHAPTER VII

Adding Functionality

Handling user input and interactions

Mobile app development is a dynamic and ever-evolving field, driven by the pursuit of creating user-friendly and engaging applications. At the heart of every successful mobile app lies the ability to handle user input and interactions seamlessly. Whether tapping a button, swiping through a gallery, or inputting text, how an app responds to user actions directly influences the user experience. This section will delve into the intricacies of handling user input and interactions in mobile app development, exploring the fundamental principles, tools, and best practices that empower developers to create intuitive and responsive applications.

User input and interactions are at the core of any interactive application. These interactions bridge the gap between users and the digital world, enabling users to communicate their intentions, navigate through content, and manipulate data. Effective handling of user input is critical for providing a fluid and enjoyable user experience, whether it's a simple tap to open an app or complex multi-touch gestures.

Consider a social media app, where users can scroll through their feeds, post updates, and engage with others. Each action involves different types of user input and interactions, from scrolling lists to typing text and tapping on buttons. The app's ability to interpret and respond to these actions defines its usability and user satisfaction.

In the mobile realm, touch is the primary mode of interaction. Users interact with mobile devices through taps, swipes, pinches, and multi-touch gestures. Developers must understand the nuances of touch interactions to create responsive and intuitive apps.

Handling touch input involves recognizing gestures and translating them into meaningful actions within the app. For instance, a swipe gesture can be used to navigate between screens or scroll through content, while a pinch gesture can zoom in or out on an image. Gesture recognition libraries and APIs provided by mobile platforms simplify detecting and responding to these interactions.

Text input is another crucial aspect of user interaction, especially in apps that involve content creation or data input. Handling text input encompasses features like virtual keyboards, text fields, and input validation.

Virtual keyboards are an integral part of text input on mobile devices. Apps need to manage keyboard visibility, ensure that it doesn't obscure important content, and provide a smooth user experience when users switch between text fields. Developers can utilize keyboard management libraries and techniques to address these challenges.

Input validation is vital for ensuring data integrity and user satisfaction. Apps should validate user inputs to prevent errors and provide feedback to users when their inputs are invalid. For instance, a password input field may require specific complexity criteria, and a date input field may require a particular format.

Buttons and controls are the most common means of user interaction in mobile apps. Users tap buttons to submit forms, confirm actions, or navigate between screens. Designing and implementing buttons and controls that are visually appealing and responsive is a fundamental aspect of user interface (UI) design.

Developers use UI frameworks and libraries to create buttons, sliders, switches, and other interactive elements that adhere to platform-specific design guidelines. These elements are styled to match the app's aesthetics and functionality while ensuring responsiveness to user taps and gestures.

Many modern mobile apps involve complex interactions and animations that go beyond simple taps and swipes. For example, gaming apps rely on multi-touch gestures for gameplay, while productivity apps use drag-and-drop interactions for tasks like reordering items in a to-do list.

Handling complex interactions often requires custom code and careful user flow and feedback consideration. Developers use animation frameworks and physics engines to create fluid and realistic animations that enhance the user experience and provide visual cues for user interactions.

Providing feedback to users is an essential part of handling interactions. Users should receive immediate and appropriate feedback when performing app actions. Feedback can include visual cues, animations, sounds, or vibrations to confirm successful actions or alert users to errors.

Moreover, accessibility considerations ensure that all users, including those with disabilities, can interact with the app effectively. Developers should follow accessibility guidelines and provide features like screen readers, voice commands, and adjustable font sizes to make apps inclusive and accessible to many users.

User input and interactions are areas where thorough testing and optimization are paramount. Usability testing, beta testing, and user feedback sessions help identify user input handling and interaction issues. Developers iterate on the app's design and functionality based on real-world user interactions, addressing pain points and enhancing the overall user experience.

In conclusion, handling user input and interactions in mobile app development is a multifaceted task that requires a deep understanding of user behavior, touch gestures, text input, UI design, and accessibility. Effective handling of these interactions is essential for creating user-friendly and engaging apps and ensuring user satisfaction and retention. By following best practices, utilizing relevant tools and frameworks, and continuously testing and optimizing user interactions, developers can create mobile apps that resonate with users and stand out in the competitive app landscape.

Creating navigation within your app

In the fast-paced world of mobile app development, creating an application is no longer just about coding features or delivering content—it's about designing a holistic user experience. Central to this experience is the concept of navigation, the art of guiding users seamlessly through an app's interface. Effective navigation design is not just a convenience; it's a cornerstone of user satisfaction and engagement. In this section, we'll delve into the intricate realm of creating navigation within your mobile app, exploring the principles, strategies, and tools that empower developers to craft navigation systems that guide users intuitively.

The significance of intuitive navigation cannot be overstated. It's the thread that ties together an app's various screens, features, and content, making them accessible and understandable to users. At its core, navigation design facilitates user interactions with the app, enabling them to find what they need and accomplish tasks effortlessly. A well-structured navigation system simplifies user journeys, reduces frustration, and ultimately enhances the overall user experience. In contrast, poor navigation can lead to confusion, abandonment, and negative app reviews.

The first principle of effective navigation design is simplicity. Keep the navigation structure straightforward and intuitive. Users should be able to grasp how to move within the app without extensive guidance or tutorials. A cluttered or overly complex navigation can overwhelm users and deter them from exploring the app further. Simplicity extends to the terminology used for navigation labels and buttons; clear and concise language aids user comprehension.

Consistency is another vital principle. Maintain uniformity in navigation elements like menus, buttons, and icons across the app. Consistency breeds predictability, helping users understand how to navigate from one screen to another. Users who rely on a consistent layout and design feel more confident and in control, resulting in a smoother user experience. Consistency also extends to the use of standard gestures and interactions that align with platform conventions, reducing cognitive load for users.

Prioritization is a critical strategy in navigation design. Place frequently used features or content within easy reach, giving them prominent positions within the navigation hierarchy. Consider the user's most common tasks and goals when arranging navigation elements. For instance, an e-commerce app might prioritize the "Shopping Cart" or "Browse Products" options, as they are central to the user's shopping experience. Prioritization ensures that the most relevant and valuable content is readily accessible.

Providing feedback is a fundamental aspect of navigation design. Users should receive immediate and appropriate feedback when they perform actions within the app. Feedback can be visual, auditory, or tactile, confirming the user's interaction. For example, changing the color of a selected menu item or button when tapped provides a clear visual cue that the selection is registered. Feedback reassures users, reducing uncertainty and improving the perceived responsiveness of the app.

Efficiency in navigation design centers on optimizing user journeys. Minimize the number of steps or interactions required for users to accomplish common tasks. Users appreciate apps that respect their time and effort. Streamlining the navigation flow ensures that users can complete tasks quickly and without unnecessary friction. Efficient navigation is often linked to effectively placing calls to action, reducing the number of taps or swipes required to access essential features.

Navigation patterns and strategies vary depending on the app's goals and content. Common navigation patterns include tab bar navigation, which uses tabs at the bottom of the screen to switch between primary sections or views, and drawer navigation, where a sliding panel reveals a menu or list of options. The choice of navigation pattern depends on factors like the number of sections or features, the app's target platform, and design preferences. For instance, tab bar navigation is suitable for apps with a few primary sections, while drawer navigation is favored in apps with extensive content or options.

The tools and frameworks available to developers are crucial in simplifying navigation design. Many mobile app development frameworks offer navigation components that streamline the creation of navigation structures. Android's Navigation Component, for example, simplifies the creation of navigation graphs and supports common navigation patterns. Similarly, React Navigation is a popular library for building navigation in React Native apps, offering a range of navigation patterns and customization options.

Additionally, native navigation APIs and libraries provided by mobile platforms enable developers to create navigation structures that adhere to platform conventions. For iOS, UIKit and SwiftUI are commonly used for building navigation interfaces, while Android offers Jetpack Navigation for streamlined navigation management. Third-party libraries and components can also enhance navigation, providing unique navigation patterns or animations.

User testing is an indispensable step in refining navigation design. Conduct usability testing with actual users to detect pain points, confusion, or bottlenecks in the navigation flow. Gather feedback on how users perceive and interact with the navigation system. Usability testing can reveal issues that may not be apparent during the development phase, helping developers fine-tune the navigation based on real-world user interactions.

Iterative design based on user feedback is a fundamental principle in navigation refinement. Developers should be prepared to adjust and improve the navigation system based on user testing and feedback. Continuous testing and iteration ensure the app's navigation remains user-centric and aligned with evolving user expectations.

In conclusion, creating navigation within your mobile app is both an art and a science. It needs a deep understanding of user behavior, interaction patterns, and design principles. Effective navigation design is not merely about moving users from one screen to another; it's about orchestrating an intuitive, efficient, and engaging user journey.

By adhering to simplicity, consistency, prioritization, feedback, and efficiency principles, developers can craft navigation systems that resonate with users and improve their overall experience. The choice of navigation patterns and tools should align with the app's objectives and design aesthetics. Through user testing and iterative refinement, developers can create mobile apps that meet user needs and provide a seamless and delightful navigation experience. In the ever-evolving landscape of mobile app development, navigation design remains vital, shaping how users interact with and perceive your app.

Implementing features and functions

Mobile app development is a dynamic and transformative field that empowers developers to bring innovative ideas to life. At the heart of every successful mobile app are its features and functions—the core functionalities that define its purpose and utility. Implementing these features and functions effectively is a crucial aspect of the development process, as it directly impacts user satisfaction and the app's success. In this section, we will explore the intricacies of implementing features and functions within your mobile app, from the initial ideation stage to the development, testing, and refinement phases, while focusing on delivering user-centric experiences.

Features and functions are the building blocks of a mobile app, the elements that users interact with to achieve specific goals and tasks. They encompass a broad spectrum of capabilities, from basic functions like displaying content and user authentication to complex features like real-time chat, augmented reality, or machine learning algorithms.

Consider a weather app; its features include displaying current weather conditions, forecasts, and interactive maps. A messaging app allows users to send messages, share media, and engage in group chats. An e-commerce app provides features for browsing products, purchasing, and managing a shopping cart. Each feature serves a unique purpose and collectively defines the app's utility and value to users.

Before implementing features and functions, beginning with a clear vision and well-defined goals for the app is essential. The ideation phase involves brainstorming ideas, conducting market research, and comprehending user needs and pain points. During this phase, the app's foundational features and functions are conceptualized.

User personas play a crucial role in shaping feature development. Identifying the target audience and their preferences informs the design and functionality choices. For example, a fitness app for beginners may focus on features like guided workouts and progress tracking, while a fitness app for experienced athletes may prioritize advanced analytics and customization options.

During the ideation phase, it's also essential to consider the competitive landscape. Analyze other apps in the same niche to identify gaps or opportunities for differentiation. Innovative and unique features can set an app apart in a crowded market.

The design phase comes into play once the core features and functions are defined. This phase involves creating wireframes, prototypes, and user interface (UI) designs that illustrate how users will interact with the app. The design should align with the app's goals and user personas, ensuring a seamless and intuitive user experience.

Consider a social media app; during the design phase, the news feed layout, the placement of buttons for liking and commenting, and the design of user profiles are all meticulously crafted to encourage user engagement and ease of use. Attention to detail in the design phase is crucial for translating ideas into a functional and visually appealing app.

In the development phase, the implementation of features and functions begins in earnest. Developers write code, create databases, and integrate third-party services to build the app's functionality. This phase requires a deep understanding of programming languages, mobile development frameworks, and best practices.

For example, a ride-sharing app relies on GPS and mapping services to calculate routes and provide real-time location tracking. Developers must implement these features while optimizing for accuracy, performance, and battery efficiency.

Moreover, the choice of development approach matters. Native development, which uses platform-specific languages and tools like Swift for iOS and Kotlin for Android, provides the best performance and access to device-specific features. Cross-platform development frameworks like React Native and Flutter offer efficiency by enabling developers to write code once and deploy it across multiple platforms.

Testing is a critical phase in the development process, as it verifies that the implemented features and functions work as intended and are free from bugs and issues. Quality assurance testing includes functional, usability, performance, and security testing.

Functional testing involves testing each feature to ensure it performs its intended task correctly. For example, in an e-commerce app, functional testing checks that users can add items to their shopping carts, proceed to checkout, and complete a purchase without errors.

Usability testing assesses the overall user experience by gathering feedback from real users. It helps identify areas where the app's features and functions may not align with user expectations or preferences. Usability testing often uncovers valuable insights for making improvements.

Performance testing evaluates the app's speed, responsiveness, and scalability. It guarantees that the app can handle a variety of scenarios, such as high user loads or slow network connections, without crashing or slowing down.

Security testing is crucial, primarily if the app handles sensitive user data or transactions. It involves identifying vulnerabilities, such as data breaches or code exploits, and implementing measures to protect user information. The app development process is rarely a one-and-done endeavor. The refinement phase involves analyzing user feedback and data, adjusting features and functions, and continuously improving the app's performance and usability.

User feedback, obtained through app reviews, ratings, and user surveys, provides valuable insights into how users perceive the app's features and functions. Addressing user feedback demonstrates a commitment to user satisfaction and helps retain and attract users.

Data analytics tools can provide valuable information on user behavior, usage patterns, and feature engagement. Developers can use this data to make data-driven decisions on which features to prioritize or enhance. Regular updates and bug fixes are essential to ensure the app remains reliable and secure. Users appreciate apps that are actively maintained and improved over time.

Throughout the process of implementing features and functions, user-centered design principles should remain at the forefront. Consider user needs, preferences, and feedback at every stage of development. Accessibility features, such as screen readers and adjustable font sizes, should be considered to ensure that the app is inclusive and accessible to a wide range of users.

Furthermore, ensure that the app's features and functions are well-documented for both users and future developers who may need to maintain or extend the app's capabilities.

Implementing features and functions within your mobile app is a multifaceted journey that transforms ideas into user-centric experiences. The process begins with imagination, where the app's core features and functions are defined based on user needs and market analysis. The design phase brings these concepts to life through wireframes and UI designs, while the development phase involves coding, database creation, and integration of services.

Testing and refinement are crucial to ensure the implemented features work as intended, provide a seamless user experience, and remain secure and reliable. User-centered design principles guide the entire process, from ideation to implementation, ensuring that the app addresses user needs and preferences.

Mobile app development is an ongoing endeavor, and continuous improvement is vital to staying competitive and meeting user expectations. By focusing on user-centric design, embracing user feedback, and maintaining a commitment to quality, developers can create mobile apps that fulfill their initial vision and provide valuable and enjoyable experiences for users worldwide.

CHAPTER VIII

Testing and Debugging

Strategies for effective testing

Effective testing is a linchpin in the world of mobile app development, where the quality, reliability, and performance of an application are paramount. It's a strategic imperative in a landscape where user expectations run high, and competition is fierce. Comprehensive testing is essential to identify and rectify issues before they reach users, thereby preserving the reputation and success of an app. This section will explore vital strategies for effective testing in mobile app development, delving into key principles, methodologies, and tools that empower developers to deliver high-quality apps that users can trust and enjoy.

The foundation of effective testing lies in comprehensive test planning. This initial phase involves defining clear objectives, identifying the scope of testing, and establishing a robust testing strategy. The test plan should outline the types of tests to be conducted, the devices and platforms to be covered, and the criteria for success. For instance, in developing a finance app, the test plan might include ensuring accurate calculations of financial transactions, validating secure data storage, and verifying compatibility with various devices and operating systems.

Effective testing encompasses various types and levels, each serving a specific purpose in evaluating the app's functionality and performance. Functional testing, for example, focuses on verifying that the app's features and functions work as intended, including tests for user interactions, data processing, and input validation. Usability testing assesses the app's user-friendliness, gathering feedback from real users to evaluate the app's ease of use, navigation, and overall user experience.

Performance testing is vital to evaluate the app's speed, responsiveness, and scalability. It includes load testing to assess the app's performance under heavy user loads and stress testing to identify potential bottlenecks or performance issues. Security testing, on the other hand, is crucial for apps handling sensitive user data or financial transactions. It identifies vulnerabilities, such as data breaches or code exploits, and ensures that the app's security measures are robust.

Test automation plays a critical role in efficient and consistent testing. Automation tools and frameworks enable the execution of repetitive tests, regression tests, and cross-platform tests with minimal manual effort. Automation is particularly valuable in mobile app development, where a wide range of device models, screen sizes, and operating systems must be tested. Real-

device testing complements emulators and simulators by replicating real-world usage scenarios. Real devices can uncover device-specific issues, such as performance disparities, hardware compatibility problems, and network conditions that may affect the app's performance. Wider test coverage is ensured through cloud-based testing platforms, which offer access to a variety of real devices for testing.

Continuous integration (CI) and continuous testing (CT) are integral to modern app development practices. CI involves the automated integration of code changes into the main codebase multiple times a day, promoting code consistency and collaboration among developers. CT extends CI by integrating automated testing into the development pipeline, identifying issues early in the development cycle to minimize the chances of introducing new bugs.

User feedback and beta testing provide real-world insights into how users interact with the app. Beta testing involves releasing a pre-release version of the app to a selected group of users who provide feedback and report issues. User feedback and beta testing can uncover issues not identified during internal testing and offer a real-world validation of the app's functionality and usability.

Clear and concise test reporting and documentation are essential for tracking test progress, identifying issues, and documenting test results. Test reports should include detailed information on test cases, test execution status, and any defects discovered. Documentation ensures that testing efforts are well-documented for future reference, aiding in regression testing and knowledge transfer among team members.

In conclusion, effective testing is the bedrock of quality in mobile app development, ensuring that apps meet user expectations for functionality, performance, usability, and security. By embracing comprehensive test planning, a holistic approach to testing types and levels, test automation, real-device testing, continuous integration and testing, user feedback, and robust reporting and documentation, developers can deliver high-quality apps that users can trust and rely on. In a dynamic and competitive app development landscape, effective testing remains the guardian of app quality, continually raising the bar for user satisfaction and success.

Debugging common issues

Mobile app development is a thrilling journey filled with innovation and creativity. Yet, it's also a journey riddled with challenges, one of the most formidable being debugging common issues. As developers strive to bring their vision to life, they often encounter bugs, errors, and glitches that can impede progress and frustrate users. This section will explore the art of debugging common issues in mobile app development, focusing on the strategies, tools, and best practices that empower developers to identify, troubleshoot, and resolve these challenges effectively.

Debugging starts with the right mindset. Developers must approach the process with patience, persistence, and a methodical mindset. Rather than relying on intuition or guesswork, they should adopt a systematic approach to diagnose and fix issues. This involves understanding the problem, replicating it consistently, and isolating its root cause.

Logging and error reporting are fundamental tools in debugging. Developers should incorporate robust logging mechanisms into their apps, which record key events, actions, and error messages. This helps in tracking down issues, as it provides a timeline of app behavior leading up to the problem. Error reporting mechanisms are equally crucial. These tools collect and transmit error information back to developers, enabling them to pinpoint issues even when users encounter them in the wild. Platforms like Firebase Crashlytics and Sentry offer powerful error tracking capabilities, providing invaluable insights into app issues.

While emulators and simulators are useful for initial testing, real-device testing is essential to bridge the gap between development and real-world usage. Real devices may uncover issues related to device-specific hardware, network conditions, or interactions that are not apparent in emulated environments. Cloud-based testing services, like AWS Device Farm and BrowserStack, provide access to a vast array of real devices for thorough testing.

Version control systems, such as Git, are pivotal in collaborative debugging. Maintaining a structured history of code changes allows developers to trace issues back to specific code commits. Additionally, code review processes help catch potential issues before they enter the codebase.

Reproducing issues consistently is essential for effective debugging. Developers should strive to create a step-by-step process that reliably recreates the problem. This not only aids in diagnosing the issue but also helps in verifying the effectiveness of potential fixes.

Remote debugging tools offer real-time insights into app behavior on a user's device. For mobile apps, remote debugging often involves connecting to a device over the internet and inspecting its runtime behavior. Popular platforms like Android Studio and Xcode provide remote debugging capabilities, allowing developers to see device logs, inspect variables, and execute code remotely.

Automated testing, including unit and UI tests, can catch issues early in development. Unit tests validate individual components of the app, while UI tests simulate user interactions to ensure that the app functions correctly. Continuous integration (CI) pipelines can run these tests automatically with each code commit, helping to identify regressions and issues before they reach users.

Mobile app development is closely tied to the evolving nature of mobile operating systems and devices. Developers must ensure their apps remain compatible with the latest OS versions and device specifications. Regularly updating app dependencies, libraries, and SDKs is essential to address compatibility issues and prevent app crashes.

User feedback is a goldmine of information for debugging. Listening to user reports, reviews, and crash reports can uncover issues developers may have missed during testing. Engaging with users and acknowledging their feedback helps resolve issues and builds trust and loyalty.

Finally, debugging is an ongoing learning process. Developers should stay informed about best practices, emerging tools, and evolving technologies. Participating in developer communities and forums can provide access to valuable insights and solutions to common issues.

In conclusion, debugging common issues is integral to the mobile app development journey. It demands a combination of technical skills, the right mindset, and a toolbox of strategies and tools. Developers who embrace a systematic approach, leverage logging and error reporting, test on real devices, collaborate effectively, reproduce issues consistently, use remote debugging, implement automated testing, prioritize version compatibility, value user feedback, and commit to continuous learning will be better equipped to navigate the debugging challenges and deliver high-quality mobile apps that users can rely on and enjoy. Debugging is not just about fixing problems; it's about refining the development process, ensuring app reliability, and ultimately delighting users with seamless, issue-free experiences.

Beta testing and user feedback

Mobile app development is an iterative journey where the quest for perfection is never-ending. As developers strive to create apps that stand out in a competitive landscape, beta testing and user feedback play pivotal roles in shaping the final product. These processes offer a unique opportunity to gather insights, identify issues, and fine-tune the user experience before launching to a broader audience. This section will explore the significance of beta testing and user feedback in mobile app development, shedding light on their benefits and best practices.

Beta testing marks a significant milestone in the development process. It involves releasing a pre-release version of the app to a select group of users, often called beta testers. These testers are typically a mix of internal team members, trusted partners, and sometimes external users who volunteer to participate. Beta testing aims to subject the app to real-world usage scenarios and collect valuable feedback.

Beta testing offers several key advantages. First, it is a powerful tool for bug identification. Beta testers can uncover issues that may have been missed during internal testing. They provide diverse testing environments, devices, and usage patterns, which can reveal device-specific bugs, performance bottlenecks, and other issues. Second, beta testing provides invaluable user feedback. Beta testers offer insights into the app's functionality, usability, and overall user experience. Their input helps identify areas for improvement, design enhancements, and feature requests. This feedback is instrumental in shaping the final version of the app to align with user preferences and expectations.

Another critical advantage of beta testing is that it replicates real-world usage scenarios. This means that developers can understand how users interact with the app daily. Real-world user observation yields insights beyond what can be replicated in a controlled setting.

Additionally, involving users in the development process can boost their satisfaction and loyalty. It shows that you value their opinions and are committed to providing a high-quality app that meets their needs. This can lead to more robust user retention rates and positive word-of-mouth recommendations, driving the app's success.

To conduct effective beta testing, planning and executing the process meticulously is crucial. Start by defining clear test objectives. Identify the specific areas or features you want testers to focus on and provide clear guidelines. Recruit diverse testers to ensure coverage of different devices, operating systems, and usage patterns. Consider involving internal team members, loyal users, and external volunteers who are enthusiastic about your app.

Establish efficient feedback mechanisms like in-app feedback forms, dedicated email addresses, or online forums. Make it easy for testers to report issues and provide suggestions. Use safe and user-friendly distribution channels, like Apple TestFlight or the Google Play Store, to launch the beta version.

Once feedback starts coming in, collect and analyze it systematically. Categorize feedback into different areas, such as bugs, usability issues, and feature requests. Prioritize issues based on severity and impact. This structured approach ensures that developers can address the most critical issues first, making the most efficient use of their time and resources.

User feedback isn't limited to the beta testing phase; it's an ongoing process throughout the app's lifecycle. Continuous iteration based on feedback is essential. Use beta testing feedback to iterate and improve the app. Regularly release updates that address reported issues and incorporate user suggestions.

Monitor app store ratings and reviews and respond to user feedback, addressing concerns and showcasing your commitment to enhancing the app. Establish feedback loops within your development team to ensure user feedback informs decision-making at every stage of development.

In conclusion, beta testing and user feedback are indispensable components of mobile app development. They provide insights into real-world usage, uncover hidden issues, and continually empower developers to enhance the user experience. By embracing a user-centered approach and making beta testing and user feedback integral parts of your development process, you can craft apps that meet user needs and thrive in a competitive market. Ultimately, the collaboration between developers and users leads to apps that users love and return to, driving success in the mobile app ecosystem.

CHAPTER IX

Publishing Your App

Preparing your app for release

The journey of mobile app development culminates in a critical phase: preparing your app for release. This phase is akin to launching a ship into the vast ocean, where careful planning and execution are vital for a smooth voyage. This section will explore key considerations and best practices in preparing your mobile app for release, from final testing and optimization to crafting compelling app store listings and marketing strategies.

Before releasing your app to the public, thorough testing and quality assurance are paramount. The app should undergo rigorous testing to identify and rectify any remaining bugs, glitches, or performance issues. Functional testing, user interface testing, and compatibility testing on various devices and operating systems are crucial to ensure a seamless user experience.

In addition to functional aspects, pay attention to performance optimization. Optimize the app's speed, responsiveness, and resource usage to provide a seamless and efficient user experience. Performance testing, including load and stress testing, can help identify bottlenecks and ensure the app can handle a significant user load.

Security is non-negotiable in today's app landscape. Ensure that your app handles user data securely and follows best practices for data protection. Encrypt sensitive data, implement secure authentication mechanisms, and conduct penetration testing to identify vulnerabilities. Adhere with data protection regulations like GDPR, CCPA, or HIPAA, depending on your app's scope and user data handling.

Crafting compelling app store listings is essential for attracting users. Your app's icon, screenshots, and app description should convey its value proposition and unique features. Optimize your app's title and description for relevant keywords to improve discoverability. Provide clear and concise information about the app's functionality, benefits, and any recent updates.

App Store Optimization (ASO) is vital to preparing your app for release. ASO involves optimizing your app's listing to rank higher in app store search results. Conduct keyword research to identify suitable keywords and phrases for which users will likely search. Use these keywords strategically in your app's title, description, and metadata. Regularly monitor your app's performance in app store rankings and make adjustments as needed to improve visibility.

Effective marketing and promotion are essential to garner attention and downloads. Develop a comprehensive marketing strategy that entails pre-launch and post-launch activities. Leverage social media, email marketing, and press releases to create buzz before your app's release. Think about having targeted advertising campaigns in order to reach the people who are most likely to use your app. Collaborate with influencers or industry partners to promote your app.

When you're confident in your app's quality, security, and ASO strategy, it's time to submit it to the app stores. Follow the submission guidelines and requirements of platforms including Apple App Store, Google Play Store, or other relevant app marketplaces. Be prepared to provide all necessary information, such as app details, screenshots, privacy policies, and developer contact information.

As your app goes live, ensure that you have robust user support and feedback channels in place. Offer users clear and accessible ways to contact your support team, report issues, or provide feedback. Promptly address user inquiries and issues to maintain a positive user experience and reputation.

After your app's release, the journey is far from over.

Monitor user feedback, app store reviews, and app analytics to gain user behavior and preferences insights. Use this data to identify areas for improvement and plan future updates. Regularly release updates to address user feedback, introduce new features, and maintain app compatibility with the latest operating systems.

In conclusion, preparing your app for release is a crucial mobile app development phase requiring meticulous planning and execution. Thorough testing, security measures, compelling app store listings, ASO strategies, marketing efforts, and user support channels are all integral to a successful launch. You can guarantee that your app reaches its target audience and offers a worthwhile and entertaining experience that continues to keep users engaged and coming back for more by adhering to best practices and paying close attention to user feedback. With careful preparation, your app is poised to set sail into the mobile app ecosystem with confidence and the potential for great success.

App store guidelines and requirements

Publishing your mobile app on app stores is the culmination of hard work and innovation. However, it's not just about creating a great app; it also involves following particular guidelines and requirements that the app store platforms have established. Whether you're targeting the

Apple App Store, Google Play Store, or any other marketplace, understanding and complying with these guidelines is essential. This section will explore the significance of app store guidelines and requirements, their impact on your app's success, and the best practices for navigating them effectively.

App store guidelines and requirements are designed to ensure app quality and a positive user experience. Submitting an app to an app store, like the Google Play or Apple App Store, comes with strict guidelines. They review apps to ensure they are free of bugs, malware, and content that violates their policies. This scrutiny aims to protect users from potentially harmful apps and maintain the overall trust and integrity of the platform. Compliance with legal and policy requirements is non-negotiable. App stores have strict rules to prevent copyright infringement, intellectual property violations, and any form of illegal content distribution. Additionally, they enforce user privacy and data protection guidelines, including handling personally identifiable information (PII). Adherence to these regulations is a prerequisite for app distribution and crucial for maintaining user trust and avoiding legal issues.

App stores often provide user interface (UI) and user experience (UX) design guidelines. These guidelines encourage consistency and usability across apps on their platforms. For example, the Apple Human Interface Guidelines (HIG) outline design principles for iOS apps, ensuring a seamless and familiar user experience for Apple device users. Adhering to these guidelines can enhance your app's usability and user satisfaction. Performance and functionality standards are essential to app quality. App stores typically require apps to be well-optimized, responsive, and free of crashes or excessive resource usage. Meeting these standards ensures that your app provides users with a smooth and reliable experience across various devices and network conditions.

App stores have content policies that dictate what can and cannot be included in an app. These policies cover many topics, from prohibited content (e.g., violence, hate speech, adult content) to app functionality (e.g., gambling, in-app purchases). It's crucial to thoroughly review and comply with these policies when developing your app to avoid rejection or removal from the store.

If your app includes monetization strategies, such as in-app purchases or advertisements, you must adhere to the app store's guidelines in this regard. This may involve disclosing pricing information, providing clear and accurate descriptions of what users are purchasing, and ensuring a fair and transparent monetization model.

Once your app is published, the commitment to app store guidelines and requirements doesn't end. App stores often require regular updates to maintain compatibility with new operating system versions and to address security vulnerabilities or bugs. Failing to keep your app up-to-date can lead to declining user ratings and losing visibility in app store search results.

Submitting your app to an app store involves a review process where the platform's team assesses your app's compliance with their guidelines and requirements. The length of time it takes to complete this process may differ based on the platform and the volume of submissions. It's essential to be patient during this phase and be prepared to address any feedback or issues raised during the review.

Navigating app store guidelines and requirements is crucial to mobile app development. It ensures that your app meets quality standards, legal and policy obligations, and provides a positive user experience. Non-compliance can result in app rejection or removal, negative user reviews, and even legal repercussions. By thoroughly understanding the guidelines, following best practices, and committing to ongoing maintenance and updates, you can position your app for success on app store platforms. Remember that compliance is not just a hurdle to overcome; it's fundamental to building and maintaining a reputable and successful mobile app in today's competitive market.

Submitting your app to app stores (Google Play Store, Apple App Store)

Bringing your mobile app to life is a remarkable achievement, but to share it with the world, you must navigate the intricate process of submitting your app to app stores. The two primary players in this realm are the Google Play and Apple App Store. Both have specific procedures, guidelines, and requirements that developers must adhere to. This section delves into the essential steps and best practices for submitting your app to these app stores, unlocking access to millions of potential users.

Before you embark on the submission journey, thorough preparations are key to success. This involves ensuring your app complies with the app store's guidelines and requirements. Review the guidelines meticulously, addressing issues such as content policies, user interface and experience standards, legal and privacy requirements, and monetization policies. Ensure your app is free from bugs, crashes, and security vulnerabilities by conducting comprehensive testing and quality assurance.

You must have a developer account with the respective platform to submit your app to either the Google Play or Apple App Store. Developer accounts typically require a one-time registration fee. Google charges a nominal fee for a Google Play Console account, while Apple's Developer Program incurs an annual fee. These accounts grant you access to the tools and resources necessary for app submission.

Crafting an appealing app store listing is vital for attracting potential users. This involves creating an engaging app icon, captivating screenshots, and a compelling description highlighting your app's unique features and benefits. Optimize your app's title and description with relevant keywords to improve discoverability in app store search results.

App Store Optimization (ASO) is the art of optimizing your app's listing to rank higher in app store search results. Conduct keyword research to identify suitable keywords and phrases users will likely search for. Strategically incorporate these keywords into your app's title, description, and metadata. Regularly monitor your app's performance in app store rankings and make adjustments as needed to enhance visibility.

Once your app is thoroughly prepared and your developer account is set up, it's time to submit it. Google Play Store provides the Google Play Console for app submission, while Apple App Store uses App Store Connect. Prepare all the required assets, including your app's binary file (APK for Android or IPA for iOS), screenshots, promotional graphics, and app store descriptions. Follow the submission guidelines, provide accurate information, and ensure your app complies with all requirements.

The Google Play and Apple App Store use a review process to evaluate submitted apps. The review process ensures that apps meet quality, security, and content standards before making them available to users. Be prepared for this review to take some time, as reviewers assess your app's functionality, usability, and adherence to guidelines. Address any issues or feedback provided during the review promptly to expedite the process.

Once your app passes the review process, it will be published on the app store platform, making it available to users for download and installation. Promote your app through marketing and advertising efforts to maximize visibility and downloads. Monitor user feedback, app reviews, and analytics to gain user behavior and preferences insights. Use this data to inform future updates and enhancements to your app.

App stores require developers to keep their apps up-to-date and compatible with the most recent operating system versions. Regularly release updates to address user feedback, fix bugs, and introduce new features. Failure to maintain your app can result in declining user ratings and visibility in app store search results.

In conclusion, submitting your app to app stores is a crucial stage in the mobile app development process. It allows you to share your creation with a vast audience and potentially achieve great success. By thoroughly preparing your app, understanding the submission guidelines, creating captivating app store listings, optimizing for search visibility, and engaging in ongoing maintenance and updates, you can maximize your app's chances of thriving in the competitive world of mobile apps. Remember that the submission process is not just a technical requirement; it's an opportunity to showcase your app to the world and make a lasting impression on users.

CHAPTER X

Marketing and Monetization

Promoting your app

Effective promotion of your mobile app is an essential next step when it is accessible in app stores. With millions of apps competing for consumers' attention in a competitive market, smart marketing is crucial to guarantee that your app gets in front of the right people and stands out. This section explores the significance of app promotion and offers insights into the strategies to help your app gain visibility, acquire users, and succeed.

App Store Optimization (ASO) is often the first step in promoting your app. ASO involves optimizing your app's app store listing to improve its discoverability in search results. To find appropriate keywords and phrases that users are likely to look for, start by performing keyword research. Integrate these keywords into your app's title, description, and metadata. Create an engaging app icon, captivating screenshots, and a compelling description highlighting your app's unique features and benefits. Regularly monitor your app's performance in app store rankings and make adjustments to enhance visibility.

Leveraging social media platforms can be a potent way to promote your app. Create dedicated social media profiles for your app and share engaging content that showcases its features, updates, and user testimonials. To reach the niche market for your app, run specific advertising campaigns on social media sites including Facebook, Instagram, Twitter, and LinkedIn. Encourage users to follow your app on social media and share their experiences, helping to generate buzz and word-of-mouth promotion.

Collaborating with influencers in your app's niche or industry can help you tap into their established audiences. Identify influencers who align with your app's target demographic and values. Contact them to explore partnership opportunities, such as sponsored posts, reviews, or giveaways. Influencer marketing can provide authentic and relatable endorsements that resonate with potential users.

You may draw people to your app and establish its authority in its field by producing insightful and useful content connected to it. Start a blog, produce how-to videos, or host webinars that address topics relevant to your app's functionality or industry. Share this content on your app's website, social media channels, and other platforms to drive organic traffic and engage potential users. Content marketing promotes your app and positions it as a valuable resource.

Consider running paid advertising campaigns within the app stores themselves. You may boost the exposure of your app to customers who are searching for keywords or similar apps by using the advertising options available in Google Play or Apple App Stores. Reaching a specific group of potential consumers with these advertisements may be an affordable option.

Leverage email marketing to engage with users who have already shown interest in your app. Build an email list by encouraging users to subscribe through your app or website. Send regular updates, newsletters, and promotional offers to keep users informed and engaged. Email marketing can help re-engage dormant users and encourage them to return to your app.

Urge customers to rate and review your app in the app stores. High ratings and favorable reviews have a big influence on how prospective consumers think and act. Respond to user reviews promptly, addressing concerns and acknowledging positive feedback. A responsive and engaged developer can create a positive impression and build user trust.

Use app analytics to monitor user behavior, track conversion rates, and identify areas for improvement. Analyze user data to refine your marketing strategies, user acquisition efforts, and app features. Continuously iterate on your app based on user feedback and data insights, striving to enhance the user experience and keep users engaged.

In conclusion, promoting your app effectively is critical to mobile app success. App Store Optimization (ASO), social media marketing, influencer collaborations, content marketing, app store advertising, email marketing, and app review management are all valuable strategies for increasing your app's visibility and acquiring users. A well-rounded promotion strategy and ongoing analytics and iteration can help your app thrive in a competitive marketplace. Remember that promoting your app is not a one-time effort but an ongoing commitment to building and maintaining a loyal user base.

App store optimization (ASO)

In the huge and rapidly growing world of mobile apps, standing out from the crowd is a formidable challenge. App Store Optimization (ASO) is a critical strategy to enhance your app's visibility, attract more users, and ultimately drive success. ASO encompasses a range of techniques and best practices to improve your app's discoverability within app stores, primarily the Apple App Store and the Google Play Store. This section delves into the significance of ASO and explores the key strategies and principles that can assist you optimize your app for better performance and growth.

In a marketplace teeming with millions of apps, app store search is often the primary method users use to discover new apps. ASO empowers developers to influence where their apps rank in search results, thereby increasing the likelihood of user discovery. Higher visibility translates to more organic installs, significantly impacting an app's success.

Keyword research is the foundation of ASO. It entails identifying the most relevant and high-traffic keywords that potential users will likely search for when looking for an app like yours. Tools and platforms like App Annie, Sensor Tower, or even the app stores themselves offer insights into popular keywords. Select a mix of competitive and long-tail keywords that accurately represent your app's features and functions.

Once you've identified your target keywords, incorporate them strategically into your app's title and description. The app title is a critical ASO element, as it carries significant weight in app store algorithms. Your title should include relevant keywords and convey the app's purpose and uniqueness. The app description allows you to elaborate on your app's features, benefits, and value proposition while incorporating keywords naturally.

Eye-catching visual assets play a crucial role in ASO. Users are drawn to visually appealing icons and screenshots. Craft an engaging app icon that is memorable and reflects your app's identity. Create captivating screenshots that showcase your app's key features and benefits. High-quality visuals can entice users to explore your app further and lead to more conversions.

Positive ratings and reviews can significantly influence users' decision to install your app. Urge satisfied purchasers to rate and review products. Respond to user comments with professionalism and empathy, regardless of whether it's positive or negative. Address concerns, fix issues, and acknowledge positive feedback. Engaging with your users improves your app's rating and demonstrates your commitment to user satisfaction.

ASO is an ongoing process. Regularly monitor your app's performance, including keyword rankings, user reviews, and conversion rates. Analyze the data to identify areas for improvement. ASO is not a one-time task; it demands constant iteration and adjustment to maintain and improve your app's position in app store search results. Consider localizing your app store listing for different regions and languages. Tailor your keywords, title, description, and visuals to cater to specific target audiences' preferences and search habits. Localization can significantly enhance your app's discoverability and appeal to a global user base.

Analyzing your competitors can offer valuable insights into ASO strategies. Examine your category's keywords, visuals, and descriptions of top-performing apps. Identify areas where your app can differentiate itself and offer a unique value proposition.

In conclusion, App Store Optimization (ASO) is a fundamental strategy for enhancing your app's discoverability and success in the highly competitive app marketplace. Effective ASO involves keyword research, optimizing app title and description, creating compelling visuals, managing ratings and reviews, continuous monitoring and iteration, localization, and competition analysis. By implementing these methods and staying committed to ASO best practices, you can increase your app's visibility, attract more users, and ultimately drive the growth and success of your mobile app. Remember that ASO is a continuous process that requires attention and adaptation to remain effective in the ever-evolving app ecosystem.

Strategies for monetizing your app

In the realm of mobile app development, creating an innovative and engaging app is just the beginning. To sustain the development and growth of your app, it's essential to employ effective monetization strategies. These strategies allow you to generate revenue from your app, whether it's a free download or a paid offering. In this section, we will explore various techniques for monetizing your app, ranging from advertising and in-app purchases to subscription models and affiliate marketing, offering insights into how you can turn your app into a profitable venture.

Advertising is one of the most popular monetization strategies for mobile apps. It involves displaying ads to users within the app and earning revenue based on ad impressions, clicks, or user interactions. Different types of ad formats include banner ads, interstitial ads, video ads, and rewarded ads. Integration with ad networks like Google AdMob, Facebook Audience Network, or AdColony allows you to tap into a vast pool of advertisers and access relevant ad inventory to your app's audience.

In-app purchases (IAPs) enable users to buy digital goods or premium features within your app. This strategy is particularly effective for freemium apps that offer a free basic version with the option to purchase upgrades or virtual items. Common examples of IAPs include unlocking additional levels in games, buying premium content in media apps, or subscribing to premium services in productivity apps. Implementing IAPs requires careful planning and user-friendly pricing to encourage conversions.

The freemium model combines free access to the app with the option for users to upgrade to a premium version for enhanced features or an ad-free experience. This approach allows you to capture a large user base while monetizing a subset of users willing to pay for additional benefits. Effective marketing and communication about the value of the premium version are key to the success of the freemium model.

Subscription-based monetization has gained popularity, especially for apps offering ongoing services, content, or premium features. Subscriptions can be delivered on a monthly, quarterly, or annual basis. This model offers a predictable and recurring revenue stream. However, it's essential to offer value to subscribers continuously and regularly update your app to retain them.

Affiliate marketing entails partnering with other businesses or products and earning a commission for driving traffic, leads, or sales to their offerings. Integrating affiliate links or promotions within your app can be a lucrative way to monetize, especially if your app has a specific niche or target audience. Affiliate marketing works well in content-centric apps, recommendation platforms, and e-commerce apps.
Collaborating with brands or businesses for sponsorships or partnerships can be a lucrative monetization strategy. Sponsorships involve featuring sponsored content or branding within your app, while partnerships may involve co-developing features or experiences with other companies. These arrangements can provide financial support and access to a broader user base.

Data monetization can be a potential revenue source for apps that collect and analyze user data. Selling aggregated and anonymized user data to businesses for market research, analytics, or advertising targeting can generate income. Prioritizing user privacy and ensuring compliance with data protection regulations is crucial when pursuing this strategy.

Crowdfunding platforms like Kickstarter or Indiegogo can be used to secure funding for app development or specific features. By presenting your app idea to potential backers, you can raise capital to cover development costs. In return, backers may receive exclusive access or rewards within the app.

In conclusion, monetizing your app is crucial to sustain its development and growth. The choice of monetization strategy depends on your app's nature, target audience, and objectives. Combining multiple strategies, such as advertising with in-app purchases or subscriptions, can provide a diversified revenue stream. Regardless of your chosen approach, prioritize user experience and transparency to ensure that monetization efforts enhance, rather than detract from, your app's value. Effective monetization strategies support your app's financial health and enable you to continue innovating and providing value to your users.

CHAPTER XI

Beyond the Basics

Advanced topics in mobile app development

Mobile app development is a continuously evolving field, driven by technological advancements, changing user preferences, and the ever-growing demand for innovative solutions. As you progress in your mobile app development journey, delving into advanced topics becomes essential to stay competitive and create apps that truly stand out. In this section, we will explore several advanced topics in mobile app development that can take your skills to the next level, from emerging technologies like augmented reality (or AR) and virtual reality (or VR) to advanced programming languages and security considerations.

AR and VR have revolutionized the mobile app landscape, offering immersive experiences and new possibilities for interaction. Developing AR and VR apps requires a deep understanding of 3D modeling, spatial mapping, and sensor integration. AR apps overlay digital content onto the real world, enhancing user experiences in gaming, navigation, and interior design. VR apps create entirely virtual environments for applications such as gaming, training simulations, and virtual tours. Exploring these technologies can open up exciting opportunities for creating cutting-edge apps.

The Internet of Things (or IoT) links everyday gadgets and devices to the internet, allowing them to collect and exchange data. Mobile apps can serve as central hubs for IoT devices, allowing users to monitor and control their smart homes, wearable devices, and industrial equipment. Developing IoT-integrated apps requires proficiency in handling real-time data streams, security protocols, and seamless device communication.

Machine learning and artificial intelligence have become integral to mobile app development, enabling personalized recommendations, natural language processing, and image recognition. Implementing ML and AI in your apps involves training models, handling large datasets, and integrating APIs and frameworks like TensorFlow and Core ML. These technologies can enhance user experiences by making apps more intelligent and adaptive.

Mastering advanced programming languages and frameworks can empower you to build more sophisticated apps. Languages like Kotlin for Android and Swift for iOS offer modern and efficient development environments. Cross-platform frameworks like Flutter and React Native enable you to create apps for numerous platforms with a single codebase. Exploring these options can streamline development and expand your app's reach.

Progressive Web Apps, also known as PWAs, are web applications that offer app-like experiences on the web. They blend the best of web and mobile app technologies, providing fast loading, offline capabilities, and push notifications. Learning to develop PWAs allows you to reach users across various devices and platforms without app store distribution.

As mobile apps handle sensitive user data, security is paramount. Advanced topics in mobile app development include implementing robust security measures, including secure authentication, data encryption, and protection against common threats like SQL injection and cross-site scripting. Staying updated on security best practices is crucial to safeguarding user information.

Cloud services and serverless computing offer scalable and cost-effective solutions for app backend development. Integrating your app with AWS, Azure, or Google Cloud can enhance performance, scalability, and data storage capabilities. Understanding serverless architecture can simplify backend development, reducing operational overhead.

Designing apps that are accessible to all users, including those with disabilities, is an advanced but essential topic. Learn about accessibility guidelines (such as WCAG) and techniques for designing apps with screen readers, voice commands, and alternative input methods in mind. Creating inclusive apps serves a broader audience and aligns with ethical and legal considerations.

In summary, there are a lot of innovative topics to research in the dynamic world of mobile app development. Whether you're interested in emerging technologies like AR and VR, want to harness the power of AI and ML, or aim to improve app security and inclusivity, advancing your skills can set you apart as a proficient and innovative developer. Continuously learning and experimenting with these advanced topics can elevate your app development capabilities, enabling you to create apps that push the boundaries of technology and offer exceptional user experiences.

Exploring additional libraries and frameworks

In the ever-evolving landscape of mobile app development, staying current with the latest tools, libraries, and frameworks is essential to create innovative and efficient applications. While the core programming languages (such as Java, Kotlin, and Swift) and primary development environments (Android Studio and Xcode) provide a solid foundation, exploring additional libraries and frameworks can significantly expand your app development capabilities. This section will delve into the significance of exploring these resources and highlight some valuable libraries and frameworks that can enhance your app development journey.

Additional libraries and frameworks can streamline development by providing pre-built components and functionalities. For example, libraries like ButterKnife (for Android) and Alamofire (for iOS) simplify common tasks such as UI element binding and network requests. These tools save development time, reduce errors, and improve code readability.

User Interface (UI) and also User Experience (UX) are pivotal in the success of mobile apps. Libraries like React Native and Flutter offer cross-platform solutions for creating visually appealing and responsive user interfaces. They allow developers to design consistent UIs across Android and iOS platforms, reducing design and development overhead.

To implement advanced features, additional libraries and frameworks are often indispensable. For example, Firebase provides a suite of tools for real-time database integration, user authentication, and cloud functions. Augmented reality (AR) apps benefit from ARKit (iOS) and ARCore (Android) for creating immersive experiences. These libraries empower developers to integrate cutting-edge features seamlessly.

Cross-platform development frameworks like React Native, Xamarin, and Flutter let developers write a code once and deploy it across several different platforms. This approach significantly reduces development time and resources, making it an attractive choice for businesses looking for a broader app reach.

Efficient data management and persistence are crucial for app performance. Libraries like Room (Android) and Core Data (iOS) provide robust local data storage and management solutions. They offer features such as data caching, querying, and synchronization, which are essential for apps dealing with substantial datasets.

Ensuring app reliability and stability requires effective testing and debugging tools. Libraries like JUnit (for Android) and XCTest (for iOS) facilitate unit testing, while tools like Detox and Espresso (for Android) and XCUITest (for iOS) enable UI testing and automation. These libraries enhance the testing process, helping developers identify and resolve issues quickly.

Efficient networking and API integration are vital for apps that rely on data from external sources. Retrofit (for Android) and Alamofire (for iOS) simplify network requests and response parsing. OAuth libraries like AppAuth (for Android and iOS) facilitate secure authentication with third-party services.

Many additional libraries and frameworks are open source, fostering vibrant developer communities. These communities offer support, documentation, and a wealth of resources for learning and troubleshooting. Engaging with these communities can accelerate your app development progress and provide valuable insights.

Exploring additional libraries and frameworks is pivotal to becoming a proficient mobile app developer. These resources offer a range of benefits, from streamlining development and enhancing UI/UX to implementing advanced features and simplifying testing and debugging. Whether you're creating a data-intensive app, an immersive AR experience, or a cross-platform solution, there's likely a library or framework that can significantly improve your development process. Embracing these tools enhances your app's functionality and positions you to stay competitive in the dynamic world of mobile app development. Continuously exploring and incorporating these resources into your toolkit is an investment that can pay dividends in the form of efficient, feature-rich, and successful mobile applications.

Keeping up with industry trends

Mobile app development is a fast-paced and dynamic field, constantly shaped by emerging technologies, evolving user behaviors, and industry trends. Staying current with these trends is not just a recommendation but a necessity for developers aiming to develop apps that resonate with users and stand out in the competitive app marketplace. This section will explore the importance of keeping up with industry trends and discuss strategies for staying informed in this ever-changing landscape.

Mobile app development is a fiercely competitive arena, with millions of applications competing for users' attention. Developers must stay attuned to the latest trends and user expectations to remain relevant and competitive. This awareness enables you to create apps that meet current demands, offer unique features, and provide a superior user experience.

The tech world constantly introduces new technologies and innovations that can transform app development. Staying informed about emerging technologies like AR, VR, IoT, AI, and blockchain allows you to harness these capabilities to create innovative and disruptive apps. For instance, AR can enhance user engagement, while AI can provide personalized recommendations.

User behavior and preferences evolve over time. Keeping up with industry trends helps you understand how users interact with apps, what features they value, and how they expect apps to perform. This knowledge informs your app design and development decisions, allowing you to align your product with user expectations.

Security and privacy concerns are becoming increasingly significant in mobile app development. New regulations, such as GDPR and CCPA, have elevated the importance of user data protection. Keeping up with data security and privacy trends is vital to ensure your apps comply with regulations and effectively protect user information.

User Interface (UI) and also User Experience (UX) design trends evolve to cater to changing user preferences and design philosophies. Staying updated on UI/UX trends can help you create visually appealing, intuitive, and user-friendly apps. Trends like minimalist design, dark mode, and gesture-based navigation influence how users interact with apps and should be considered in your development process.

The demand for cross-platform app development continues to rise. Frameworks like React Native, Flutter, and Xamarin offer cost-effective and time-efficient solutions for building apps that work seamlessly across multiple platforms. Staying current with cross-platform development trends enables you to leverage these tools effectively.

Staying updated with industry trends necessitates continuous learning. Consider taking online courses, attending workshops, participating in developer communities, and reading industry publications and blogs. Networking with other developers and attending conferences and meetups can also provide valuable insights and connect you with the broader developer community.

In the ever-evolving tech landscape, adaptability is essential for career longevity. Keeping up with industry trends enhances your skills and future-proofs your career. It positions you as a knowledgeable and agile developer who can embrace new challenges and opportunities as they arise.

Keeping up with industry trends is not just a professional responsibility; it's a strategic advantage in mobile app development. Staying informed about emerging technologies, user behavior, security, design, and development practices ensures your apps remain competitive, relevant, and user-centric. In a field where innovation is the norm, continuous learning and adaptation are keys to success. By embracing industry trends, you empower yourself to create apps that meet current demands and anticipate future needs, setting the stage for long-term success in the dynamic world of mobile app development.

CHAPTER XII

Troubleshooting and Common Pitfalls

Addressing common issues and errors

Mobile app development is a complex process that often involves a multitude of challenges and potential pitfalls. From coding errors to performance issues, developers frequently encounter a range of common issues that can impact an app's functionality, user experience, and overall success. In this section, we will explore some of these common issues and errors in mobile app development and discuss strategies for effectively addressing and mitigating them.

Bugs and crashes are among the most prevalent issues in app development. They can result from coding errors, memory leaks, or compatibility issues with oarticular devices or operating system versions. To address these issues, rigorous testing is essential. Implement unit testing, integration testing, and UI testing to identify and solve bugs before they reach users. Utilize crash reporting tools to monitor and analyze app crashes in real-time, enabling rapid resolution.

Performance issues like slow loading times and laggy interactions can severely impact user satisfaction. Common culprits include inefficient code, excessive memory usage, and poor network optimization. Profiling and benchmarking tools can help identify performance bottlenecks. Employ techniques like lazy loading, background processing, and data caching to optimize app performance. Regularly test your app on various devices to ensure smooth operation across different hardware configurations.

The diversity of Android and iOS devices and their varying operating system versions pose compatibility challenges. Apps that work flawlessly on one device may encounter issues on another. To mitigate compatibility challenges, conduct thorough device testing, considering different screen sizes, resolutions, and OS versions. Implement responsive design principles and utilize adaptive layouts to ensure consistent user experiences across devices.

A subpar user experience can lead to high abandonment rates and negative reviews. Common UX issues include confusing navigation, cluttered interfaces, and inconsistent design. User testing and usability studies can uncover usability issues early in development. Emphasize intuitive navigation, clear visual hierarchy, and adherence to platform-specific design guidelines to enhance UX.

Security breaches can have severe consequences, including data breaches and user privacy violations. Common security vulnerabilities include inadequate data encryption, insufficient input validation, and improper session management. Stay informed about the most recent security best practices and guidelines. Utilize encryption protocols, implement authentication and authorization mechanisms, and conduct security audits and penetration testing to identify and address vulnerabilities proactively.

Inadequate testing can result in overlooked issues and an unreliable app. Comprehensive testing should cover functional, performance, security, and usability aspects. Implement automated testing scripts for repetitive tasks and regression testing. Conduct beta testing with diverse users to identify real-world issues and gather feedback.

Apps that rely on network connectivity can encounter problems like slow data loading, failed requests, or excessive data consumption. Implement efficient network request handling, including error handling and retries. Utilize background synchronization and data prefetching to enhance user experiences, especially in low-bandwidth or offline scenarios.

Insufficient documentation can hinder collaboration among developers and make it challenging to maintain and troubleshoot an app. Maintain comprehensive documentation that covers code architecture, APIs, libraries, and third-party integrations. Document coding standards and best practices to ensure consistency in the development team.

Addressing common issues and errors in mobile app development is integral to delivering a reliable, high-quality app that meets user expectations. Effective strategies include thorough testing, performance optimization, compatibility testing, UX design principles, security measures, and comprehensive documentation. By recognizing these common challenges and proactively addressing them, developers can create apps that function flawlessly and provide exceptional user experiences, fostering user loyalty and success in the competitive mobile app landscape.

Tips for maintaining and updating your app

The journey is far from over once you've successfully launched your mobile app. In fact, it's just the beginning. Mobile app development is an ongoing process that requires continuous effort and attention to ensure your app remains relevant, reliable, and competitive in the ever-evolving mobile landscape. This section will explore essential tips for maintaining and updating your app, covering aspects such as bug fixes, performance enhancements, feature updates, user feedback, and the importance of a well-planned maintenance strategy.

Bugs and issues are inevitable in app development. User feedback, crash reports, and monitoring tools can help you identify and prioritize these issues. Regularly release bug fixes to address them promptly. Keeping your app bug-free improves user satisfaction and prevents negative reviews and uninstalls.

Performance issues can drive users away from your app. Use profiling and benchmarking tools to identify performance bottlenecks. Optimize code, implement efficient algorithms, and address memory leaks. Regularly test your app on various devices and network conditions to ensure optimal performance across different scenarios.

User feedback is invaluable for app improvement. Encourage users to provide feedback and reviews. Listen to their suggestions and criticisms. Use this feedback to prioritize feature enhancements and bug fixes. Engaging with your user community improves your app and fosters loyalty.

App security is a paramount concern. Stay informed about the most current security threats and vulnerabilities. Regularly update libraries and dependencies to patch security issues. Implement secure coding practices and conduct regular security audits to protect user data and privacy.

Operating system updates can introduce compatibility issues. Stay informed about upcoming OS updates and beta releases. Test your app on new OS versions and ensure it remains functional and visually consistent. Timely updates prevent your app from becoming obsolete.

Keeping your app fresh and engaging is essential for user retention. Plan feature updates and enhancements based on user feedback and emerging trends. Introduce new functionalities, improve existing features, and innovate to meet evolving user needs. Feature updates can rekindle user interest and attract new users.

Experimentation and A/B testing can help you make data-driven decisions. Test different variations of features, UI elements, and user flows to determine what resonates most with users. A/B testing enables you to refine your app based on user preferences and behaviors.

For content-driven apps, such as news, e-commerce, or social media, regularly update the content to keep users engaged. Fresh content retains users' interest and encourages them to revisit your app. Implement content management systems to streamline content updates.

A well-maintained app requires well-maintained documentation. Keep your codebase, APIs, and libraries well-documented. Document coding standards and best practices to ensure continuity within your development team. Well-organized documentation makes it easier to onboard new team members and troubleshoot issues. Before releasing updates, rigorously test your app. Conduct thorough regression testing to ensure new features or bug fixes do not introduce new issues. Utilize automated testing scripts and real user testing to validate changes. Test across various devices and operating system versions to catch compatibility issues. Maintaining and updating your app is an ongoing commitment to excellence. It's a journey that involves addressing bugs, optimizing performance, integrating user feedback, ensuring security, staying compatible with OS versions, adding new features, and keeping content fresh. By following these tips and developing a well-planned maintenance strategy, you can retain existing users and attract new ones, ensuring that your app continues to thrive in the dynamic and competitive mobile app landscape. Remember, a well-maintained app is a testament to your commitment to delivering value to users and sustaining success in the long run.

Resources for getting help

Mobile app development is a rewarding but complex journey, and developers often encounter challenges that require assistance to overcome. Whether you're a beginner learning the ropes or an experienced developer facing a perplexing problem, knowing where to find help is crucial. In this section, we'll explore a range of resources available to developers seeking assistance, from online communities and forums to documentation, developer support, and mentorship programs.

Online communities and developer forums are vibrant spaces where developers of all levels congregate to seek and offer help. Platforms like Stack Overflow, GitHub Discussions, Reddit's r/programming, and specialized forums for Android, iOS, and other platforms are excellent places to post questions, share insights, and learn from others' experiences. When seeking help, provide clear and concise information about your issue, along with relevant code snippets or error messages, to receive effective responses.

The official documentation for development platforms, programming languages, and libraries is a treasure trove of information. It provides in-depth explanations, tutorials, code samples, and API references. Learning to navigate and utilize documentation effectively is a valuable skill. Whenever you encounter a problem or question, consult the relevant documentation first—it often holds the answer you need.

Major platform providers like Apple and Google offer developer support programs. These programs provide direct access to technical support teams who can help you resolve platform-specific issues or challenges. While some support options may be free, others require a paid subscription or membership in the platform's developer program. Consider these options when facing platform-specific issues that standard online communities may not address.

Mentorship can significantly accelerate your learning and problem-solving skills. Finding an experienced mentor in the mobile app development field can provide guidance, share insights, and offer constructive feedback. Peer review is another valuable practice—collaborate with fellow developers to review code, share knowledge, and learn from each other's experiences. Joining development teams or participating in open-source projects can facilitate mentorship and peer review opportunities.

Online learning platforms like Udemy, Coursera, edX, and many others offer paid courses and bootcamps on mobile app development. These courses often include a structured curriculum, hands-on projects, and access to instructors or mentors. Paid courses can be a worthwhile investment, especially if you want to acquire in-depth knowledge or specific skills in a guided environment.

Attending developer conferences, meetups, and hackathons provides networking opportunities and access to professionals in the field. Many conferences offer sessions where developers can seek advice, learn about best practices, and gain insights into the latest trends and technologies. Networking with fellow attendees can lead to valuable connections and resources.

Online code repositories like GitHub and GitLab host millions of open-source projects. These repositories provide access to a vast codebase and offer issue tracking, discussions, and collaboration features. Exploring open-source projects related to your app's domain can provide solutions to common problems and inspire best practices.

Numerous educational websites and blogs focus on mobile app development. Websites like RayWenderlich, Android Developers, iOS Dev Weekly, and Medium host articles, tutorials, and resources created by experienced developers. Regularly following these websites and blogs can inform you about industry trends and best practices.

In conclusion, the mobile app development journey is filled with challenges and learning opportunities. Knowing where to find help and resources is valuable for developers at all levels. You can navigate challenges effectively, learn from others, and stay up-to-date by leveraging online communities, official documentation, developer support programs, mentorship, paid courses, conferences, code repositories, and educational websites. Don't hesitate to seek assistance when needed; after all, the collaborative nature of the developer community is one of its greatest strengths.

CONCLUSION

Summarizing key takeaways

Throughout this book, "Mobile App Development: Mobile App Development 101- A Step-by-Step Guide for Beginners," we've embarked on a comprehensive journey through the world of mobile app development. We've explored the essential concepts, tools, and strategies that will empower you to create successful and user-friendly mobile applications. Let's summarize the key takeaways that can guide you on your way to becoming a proficient mobile app developer:

Understanding the Fundamentals: We began by defining mobile app development and highlighting its importance in today's digital landscape. We discussed the various platforms, including iOS, Android, and cross-platform development, helping you make informed choices about where to start.

Setting Up Your Development Environment: You learned about the significance of choosing the right operating system (Windows, macOS, Linux) and how to install the necessary software, including Integrated Development Environments (IDEs) and Software Development Kits (SDKs).

Mastering Programming Languages: We delved into the primary programming languages for mobile app development, including Java, Kotlin, Swift, and others. Understanding their nuances is crucial for effective coding.

Grasping Essential Concepts: You gained insights into essential programming concepts, including variables, data types, and basic syntax. These fundamentals serve as the building blocks of your coding journey.

Navigating Control Structures: We explored control structures like if statements and loops, which are vital for creating logic in your apps. Mastery of these structures is essential to control app behavior effectively.

Building Your First App: We guided you through creating a "Hello World" app, a fundamental milestone in any developer's journey. This hands-on experience helped you become familiar with the development process.

Understanding App Structure: You explored the structure of mobile apps, both from a user interface (UI) and code perspective. Properly organizing your app's components ensures clarity and maintainability.

Testing and Deployment: We discussed running your app on an emulator or device for testing, a critical step before releasing it to the public. Ensuring your app functions as expected is essential for user satisfaction.

Prioritizing User Experience: Principles of good UI/UX design were emphasized to create apps that users find intuitive and visually appealing. User-centric design is key to success.

Designing Layouts and Elements: You learned how to design app layouts and elements, considering usability, accessibility, and aesthetics. Effective UI design can enhance user engagement.

Storing and Retrieving Data: Understanding data storage and retrieval methods is crucial for creating data-driven apps. Databases, APIs, and data models enable your app to interact with and manipulate data effectively.

Handling User Input: You gained insights into managing user input and interactions, including touch events, gestures, and form input. User-friendly input mechanisms are vital for app usability.

Creating Navigation: We discussed techniques for creating navigation within your app, allowing users to move seamlessly between different screens and sections. Effective navigation enhances user experience.

Implementing Features and Functions: You explored the process of adding features and functions to your app, from creating login systems to integrating APIs and implementing advanced functionalities.

Testing and Debugging: We highlighted strategies for effective testing and debugging, including unit testing, UI testing, and debugging common issues. Ensuring your app's stability is critical.

Implementing Data Storage: You learned about data storage options, from local storage to cloud-based solutions. Choosing the right storage method depends on your app's needs and scalability requirements.

Beta Testing and User Feedback: We discussed the importance of beta testing and collecting user feedback to identify issues and make improvements. User input is invaluable for refining your app.

Preparing for Release: Preparing your app for release involves meeting app store guidelines and requirements, optimizing app store listings, and conducting final testing to ensure a smooth launch.

Monetization Strategies: You explored strategies for monetizing your app, including in-app advertising, freemium models, and subscription plans. Effective monetization can help you generate revenue from your app.

Exploring Advanced Topics: Finally, we touched on advanced topics like the augmented reality (AR), virtual reality (VR), machine learning (ML), and cross-platform development. These areas offer opportunities for innovation and differentiation.

In conclusion, this book has provided you with a comprehensive foundation in mobile app development. While it's just the beginning of your journey, you now have the knowledge and tools to create, test, and deploy mobile apps. Remember that mobile app development is a continuously evolving field, and staying informed about industry trends and leveraging available resources will be key to your ongoing success. Embrace a lifelong learning mindset, seek assistance when needed, and refine your skills to build impactful and successful mobile applications.

Encouragement for readers to continue learning and building their skills

As you've journeyed through the pages of this book, "Mobile App Development: Mobile App Development 101- A Step-by-Step Guide for Beginners," you've gained valuable insights into the exciting world of mobile app development. You've explored the fundamentals, learned to navigate the intricacies of various platforms, and acquired the tools and knowledge necessary to create your own mobile applications. As we conclude this guide, it's crucial to emphasize the importance of ongoing learning and skill-building in this dynamic field.

Mobile app development is a constantly evolving discipline, shaped by emerging technologies, changing user expectations, and industry trends. Therefore, your journey as a developer is far from complete—it's a continuous process of growth and adaptation. Here are some words of encouragement to inspire you to continue your learning and skill-building endeavors:

Embrace Lifelong Learning: The tech industry is characterized by its rapid pace of change. New programming languages, frameworks, and tools emerge regularly, offering exciting opportunities for innovation. Embrace the mindset of lifelong learning, stay curious, and be open to exploring new technologies and techniques.

Stay Informed: Stay up-to-date with industry news, blogs, and forums. Follow influential figures in the field on social media platforms like Twitter and LinkedIn. Subscribe to newsletters, podcasts, and YouTube channels dedicated to mobile app development. Being informed about the latest trends and best practices is essential for staying competitive.

Expand Your Horizons: Don't limit yourself to what you've learned. Consider delving into advanced topics such as augmented reality (AR), virtual reality (VR), machine learning (ML), and Internet of Things (IoT). These domains present exciting possibilities for app innovation and differentiation.

Join Developer Communities: Engage with developer communities, both online and offline. Participate in forums, attend meetups, and collaborate on open-source projects. Building connections with fellow developers can provide support, inspiration, and opportunities for collaboration and mentorship.

Challenge Yourself: Set goals and challenges to push your boundaries. Take on projects requiring learning new skills or tackling complex problems. Personal projects, hackathons, or contributing to open-source projects can be excellent growth opportunities.

Seek Feedback: Don't hesitate to seek feedback on your work. User feedback, code reviews, and peer evaluations are invaluable for improvement. Constructive criticism helps you determine areas for growth and refinement.

Focus on Problem Solving: Mobile app development is not just about coding but solving real-world problems. Keep the end user in mind and aim to create apps that provide value and solve challenges in a user-friendly way.

Maintain a Portfolio: Document your projects, big or small, in a portfolio. Share your work on platforms like GitHub or personal websites. A well-organized portfolio showcases your skills and serves as a testament to your dedication and progress.

Take Breaks and Stay Balanced: While it's essential to keep learning and improving, remember to strike a balance. Taking breaks, practicing self-care, and pursuing hobbies outside of development can help prevent burnout and foster creativity.

Stay Resilient: Challenges and setbacks are part of any learning journey. Don't be discouraged by difficult moments or initial failures. Resilience and perseverance will serve you well in your development career.

In summary, mobile app development is a rewarding and ever-evolving field that provides endless opportunities for growth and innovation. As you continue your journey, remember that every developer, regardless of experience, was once a beginner. You'll evolve into a proficient and accomplished mobile app developer through continuous learning, experimentation, and determination. So, keep exploring, building, and pushing the boundaries of what you can create. The key to unlocking the tremendous opportunities that lie ahead will be your unwavering commitment to education and growth.

Thank you for buying and reading/ listening to our book. If you found this book useful/ helpful please take a few minutes and leave a review on the platform where you purchased our book. Your feedback matters greatly to us.

www.ingramcontent.com/pod-product-compliance
Lightning Source LLC
Chambersburg PA
CBHW071522150726
48000CB00002B/641